66th Porcelain Enamel Institute Technical Forum

Ceramic Engineering & Science Proceedings Volume 25, Issue 5, 2004

Series Editor: Greg Geiger
Director, Technical Publications: Mark Mecklenborg

Editorial and Circulation Offices
PO Box 6136
Westerville, Ohio 43086-6136

Contact Information
Editorial: (614) 794-5858
Customer Service: (614) 794-5890
Fax: (614) 794-5892
E-Mail: info@ceramics.org
Website: www.ceramics.org/cesp

Ceramic Engineering & Science Proceedings (CESP) (ISSN 0196-6219) is published five times a year by The American Ceramic Society, PO Box 6136, Westerville, Ohio 43086-6136; www.ceramics.org. Periodicals postage paid at Westerville, Ohio, and additional mailing offices.
The American Ceramic Society assumes no responsibility for the statements and opinions advanced by the contributors to its publications. Papers for this issue were submitted as camera-ready by the authors. Any errors or omissions are the responsibility of the authors.
Change of Address: Please send address changes to *Ceramic Engineering and Science Proceedings*, PO Box 6136, Westerville, Ohio 43086-6136, or by e-mail to info@ceramics.org.
Subscription rates: One year $220 (ACerS member $176) in North America. Add $40 for subscriptions outside North America. In Canada, add GST (registration number R123994618).
Single Issues: Single issues may be purchased online at www.ceramics.org or by calling Customer Service at (614) 794-5890.
Back Issues: When available, back issues may be purchased online at www.ceramics.org or by calling Customer Service at (614) 794-5890.

Indexing: An index of each issue appears at www.ceramics.org/ctindex.asp.
Contributors: Each issue contains a collection of technical papers in a general area of interest. These papers are of practical value for the ceramic industries and the general public. The issues are based on the proceedings of a conference. Both The American Ceramic Society and non-Society conferences provide these technical papers. Each issue is organized by an editor, who selects and edits material from the conference proceedings. The opinions expressed are entirely those of the presenters. There is no other review prior to publication. Author guidelines are available on request.
Postmaster: Please send address changes to *Ceramic Engineering and Science Proceedings*, P.O. Box 6136, Westerville, Ohio 43086-6136. Form 3579 requested.

Ceramic Engineering & Science Proceedings Volume 25, Issue 5, 2004

66th Porcelain Enamel Institute Technical Forum

Steve Kilczewski
Conference Director

Holger Evele
Assistant Conference Director

William D. Faust
Editor

April 26–29, 2004
Nashville, Tennessee

Published by
The American Ceramic Society
735 Ceramic Place
Westerville, OH 43081

ISSN 0196-6219

Contents

66th Porcelain Enamel Institute Technical Forum

Statement of Ownership, Management, and Circulation

Publication Title: CERAMIC ENGINEERING AND SCIENCE PROCEEDINGS. Publication number 0196-6219. Filing date: October 1, 2004. Published 5 times (irregular) per year. Annual subscription price: $220.00. Office of Publication: 735 Ceramic Place, Westerville, Franklin County, Ohio 43081-8720. Publisher: Mark Mecklenborg, 735 Ceramic Place, Westerville, OH 43081-8720. Editor: N/A. Managing Editor: Bill Jones, 735 Ceramic Place, Westerville, Ohio 43081-8720. Owner: The American Ceramic Society, 735 Ceramic Place, Westerville, Ohio 43081-8720. Stockholders owning or holding 1% or more of total amount of stock: none. The known bondholders, mortgagees, and other security holders owning or holding 1% or more of total amount of bonds, mortgages or other securities: none.

The average number of copies each issue during the preceding 12 months:
- a) Total number of copies (net press run) . . . 1,032
- b) Paid/requested circulation
  - 1) Paid/requested mail subscriptions . . . 471
  - 2) Sales through dealers and carriers, street vendors, counter sales and other nonUSPS paid distribution . . . 372
- c) Total paid/requested circulation . . . 843
- d) Free distribution by mail, samples, complimentary and other free copies . . . 43
- e) Free distribution outside the mail . . . 9
- f) Total free distribution . . . 52
- g) Total distribution . . . 895
- h) Copies not distributed . . . 138
- i) Total . . . 1,032

Percent paid /requested circulation . . . 94%

The actual number of copies for single issue nearest filing date:
- a) Total number of copies (net press run) . . . 998
- b) Paid/requested circulation
  - 1) Paid/requested mail subscriptions . . . 581
  - 2) Sales through dealers and carriers, street vendors, counter sales and other nonUSPS paid distribution . . . 238
- c) Total paid/requested circulation . . . 819
- d) Free distribution by mail, samples, complimentary and other free copies . . . 41
- e) Free distribution outside the mail . . . 9
- f) Total free distribution . . . 50
- g) Total distribution . . . 869
- h) Copies not distributed . . . 129
- i) Total . . . 998

Percent paid /requested circulation . . . 94%

I certify that the statements made by me above are correct and complete
Mark Mecklenborg, Publisher

Foreword

We are pleased to deliver to you the proceedings of the 66th Annual PEI Technical Forum. It represents the successful completion of a year's worth of planning and preparation, culminating in three days of meetings and seminars at the Sheraton Nashville Downtown Hotel in Nashville, Tennessee on April 26-29, 2004. As you receive these proceedings, work is already progressing on the 67th Technical Forum, to be held on May 2-5, 2005 in Nashville, Tennessee, this year at the Doubletree Hotel, Nashville.

I would like to thank my vice-chairman Holger Evele (Ferro), as well as the members of the Technical Forum Committee for their time, efforts, and supportive endeavors on behalf of this year's forum. The success of the forum is directly attributable to their contributions. Each year we strive to uphold the tradition of offering information that has both useful and practical applications for our industry. I believe that we have accomplished that challenge, and have set the bar even higher for next year.

The Back-to-Basics Seminar continues to be a key element of the Technical Forum. Our gratitude goes to Holger Evele and Anthony Mazzuca (Pemco) for orchestrating another outstanding seminar, and also to the faculty staff that implemented the program. This seminar continues to be a well-attended favorite, attracting both newcomers to the porcelain enamel industry, as well as seasoned veterans. Again, thanks to all involved.

The thanks of the entire Committee goes out to this year's excellent group of speakers who provided us with information on the latest in materials and equipment used in the porcelain enameling process. We are grateful to them for their time and efforts in researching, preparing and presenting their informative papers. Additional thanks go to those suppliers who participated and supported the ever-popular Supplier's Mart. Our final thanks go to all who attended and participated in the 66th Annual PEI Technical Forum.

Holger and I will again head up the planning efforts for the 2005 Technical Forum. Following my tenure, Holger will assume the chairmanship, working with a new vice-chairman for the years of 2006 and 2007. We can look ahead to the future with assurance that the Technical Forum will continue to highlight the best that our industry has to offer. Please mark your calendar for next year's meeting May 2-5, 2005 in Nashville. It promises to be informative and worthwhile.

Steve Kilczewski, Pemco Corporation
Chairman 2004 PEI Technical Forum Committee

A.I. Andrews Memorial Lecture: Basic Steelmaking

Nancy Keller and Matt Greenwood
AK Steel Corporation

Abstract

Steel has been fundamental to the commercial enameling process since the early 1900s. During this time, the steelmaking processes have changed as well the composition and quality of steels. Over the years, the gauge of steel that is enameled has become thinner, and, in turn, the enamels also have become thinner. This has reduced the weight and reduced or maintained a lower price for appliances. The steel industry also has consolidated over the past several decades to fewer suppliers. A similar trend has occurred in the enameling industry with fewer frit suppliers and fewer but higher-volume appliance producers.

Today, the basic steelmaking process utilizes basic oxygen furnaces to refine the steel from the blast furnaces and continuous casters to form continuous strips of steel that are subsequently cut and rolled for a variety of

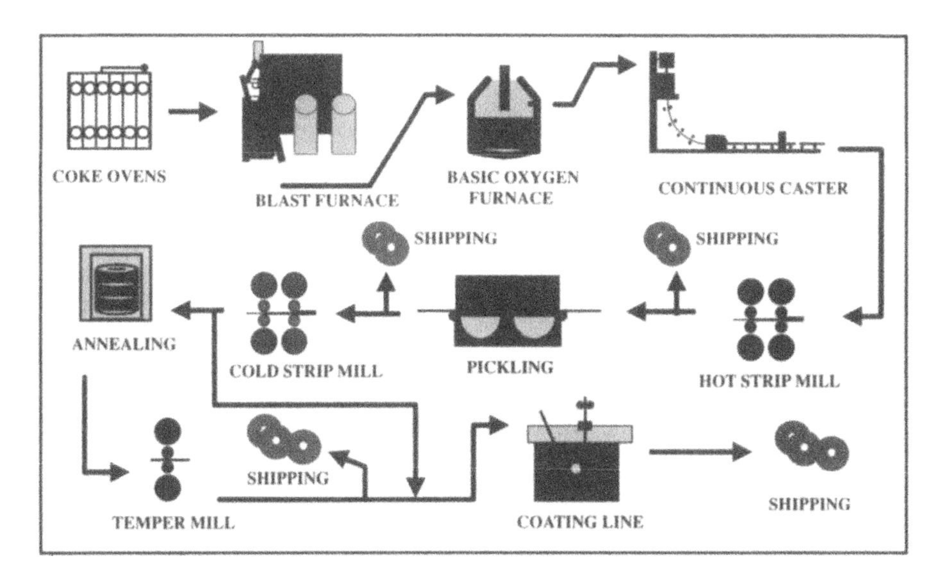

Basic steelmaking flow.

purposes. The continuously cast steels are cleaner than their ingot cast predecessors and have more tightly controlled chemistries. The grades of steel available for enameling are typically low-carbon steel (0.003–0.008% carbon) (ASTM Type I), enameling iron equivalent continuously cast steel (0.03–0.05% carbon) (ASTM Type II), and interstitial free steel (~0.10 % carbon) (ASTM Type III).

The starting point of steel is the combining of iron ore with limestone, coke, and hot air in a blast furnace. The "pig iron" produced is then transferred to a basic oxygen furnace for refining and adjustment of the chemistry to create the various steel grades before it goes to the continuous caster. Oxygen is injected into the steel to oxidize impurities. When the basic oxygen furnace is tapped, ladle additions of various metals are made. Vacuum degassing may be done to further refine the steel. The molten metal is then poured into a tundish that holds the molten steel before it is molded into strips, which may be 9 in. thick and 30 in. wide, in the casters. The strip is then cut into slabs of ~30 ft in length for subsequent rolling in a hot strip mill. Once rolled to a desired thickness, the steel is pickled to remove surface scale and impurities and then rolled again through a cold strip mill to achieve the final gauge thickness and surface properties. The steels are box annealed to relieve stresses and, with Type I, reduce the carbon content to less than 0.008% and typically less than 0.004%. After the coils are annealed, they are sent through temper mills. The tempering process retards aging, improves the flatness of the steel, and imparts a uniform surface texture.

Blast Furnace

Recipe

IRON ORE +

LIMESTONE +

COKE +

HOT AIR

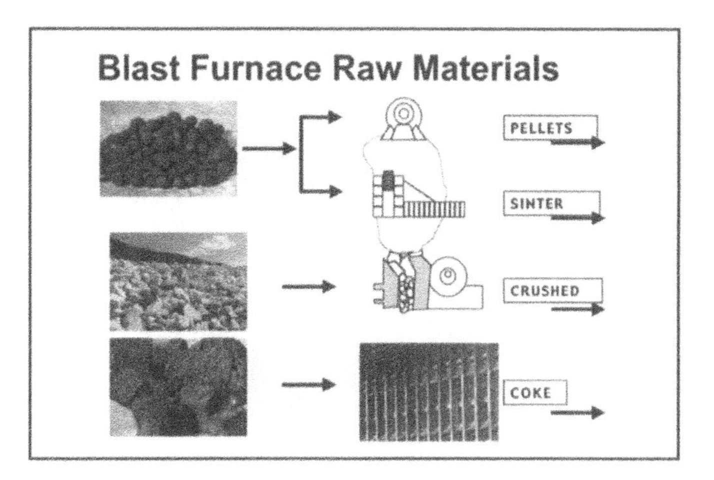

Coke Plant

The Coke Plant produces metallurgical coke for the Blast Furnace operation. The coke acts as both a fuel and a support in iron ore reduction in the Blast Furnace.

A mixture of coal grades is heated in the absence of air to drive off impurities, leaving homogeneous carbon. This process is called destructive distillation.

By-products of the coke operation include coke oven gas, tar, fertilizers, Naphtha, sulfuric acid, Creosote, Benzene, and Toluene. These are processed and collected at a recycle plant.

Blast Furnace

Blast Furnace

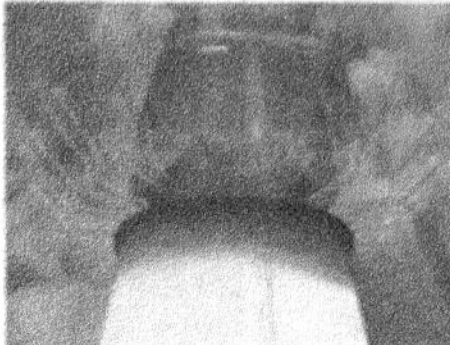

Blast Furnace

The Blast Furnace supplies molten pig iron to the Basic Oxygen Furnace. Iron ore/sinter (iron-bearing materials), coke (fuel), and limestone (flux) are charged into the top of the furnace.

Heated air is blown in at the bottom of the furnace which burns part of the fuel to produce heat for the chemical reactions involved and for melting the iron. The fuel and combustion gases remove the oxygen from the iron ore.

Molten pig iron is tapped into "torpedo" cars which transport the pig iron to the BOF. This pig iron contains approximately 4% carbon and is high in impurities such as manganese, sulfur, phosphorus, and silicon.

It takes 2 tons of air and 1.7 tons of ore to produce 1 ton of pig iron.

Blast Furnace

COKE (carbon) + OXYGEN ⟶ CARBON MONOXIDE + HEAT

$C + O_2 \longrightarrow CO$ + **HEAT**

IRON ORE + CARBON MONOXIDE ⟶ IRON + CARBON DIOXIDE

$FeO + CO \longrightarrow Fe + CO_2$

LIMESTONE "SLAG" - LIQUID BLANKET - FLOATS ON IRON BATH

Basic Oxygen Furnace

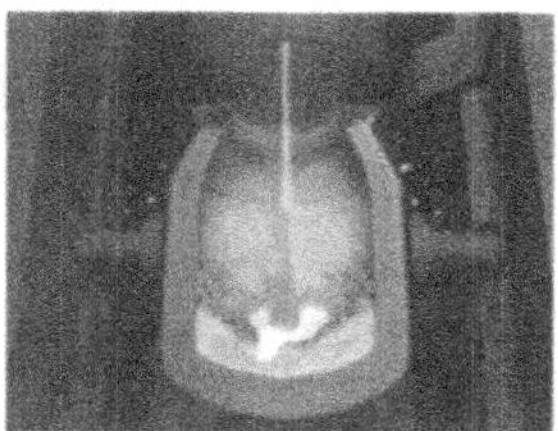

Basic Oxygen Furnace

The BOF provides molten steel to the vacuum degasser or continuous caster.

Molten pig iron and scrap (~25%) are charged into the vessel (230 Tons/Charge). A lance injects high purity oxygen into the vessel, causing a violent chemical reaction as impurities are oxidized. Silicon, manganese, and phosphorus are oxidized and float into the slag layer covering the molten steel. The carbon forms carbon monoxide gas. The oxidation of the carbon in the bath creates heat which fuels the reaction.

When the proper bath chemistry and temperature are reached, the heat is tapped into a ladle. Ladle additions (Aluminum, Columbium, Titanium) can now be made. Further refining may also be performed at the vacuum degasser where the carbon level can be further reduced. The ladle is then taken to the continuous caster.

Basic Oxygen Furnace

OXYGEN + CARBON → CO↑ + HEAT

OXYGEN + MANGANESE → MnO (SLAG)

OXYGEN + PHOSPHORUS → P_2O_5 (SLAG)

OXYGEN + SILICON → SiO_2 (SLAG)

Vacuum Degasser

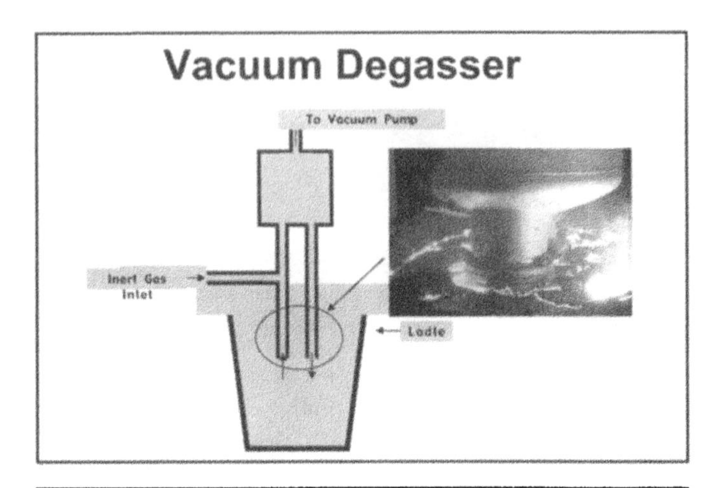

Vacuum Degassing

The Vacuum Degasser supplies a further refined molten steel to the continuous caster.

Carbon can be removed along with dissolved atmospheric gases at the vacuum degasser.

To achieve the low carbon contents required for interstitial-free steels and ultra-deep drawing steel, the steel must be vacuum degassed.

The Middletown Works has a R-H type degasser.

A vacuum unit with two snorkels is submerged into the molten steel ladle from the BOF. A vacuum is pumped and argon is entered into one snorkel. This causes the molten steel to be continuously circulated through the vacuum chamber.

Continuous Caster

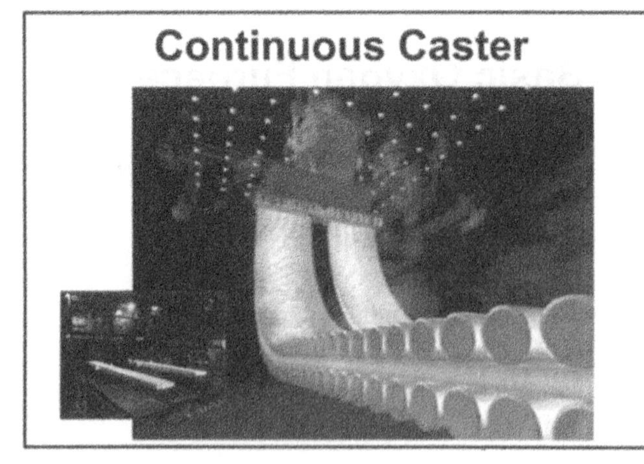

Continuous Caster

The Continuous Caster supplies steel slabs to the Hot Strip Mill.

Killed ladles of molten steel from the BOF are tapped into a tundish. The tundish is a large trough which holds the molten steel and funnels the steel into vertical slab molds. The water cooled molds oscillate and produce 9" thick slabs. Slabs are torch cut to 30+ feet long.

Hot Strip Mill

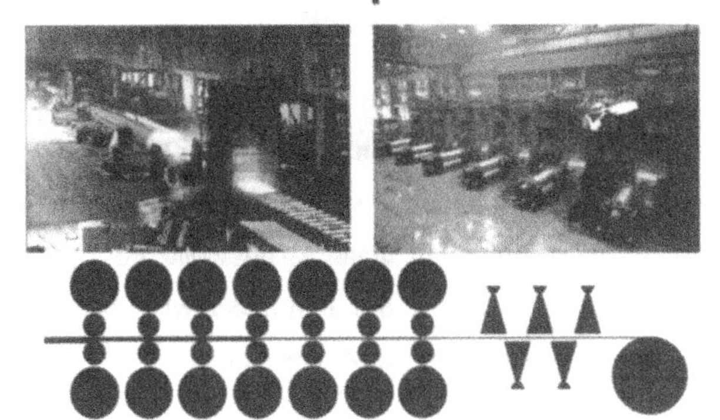

Hot Strip Mill

The Hot Strip Mill supplies hot bands to the Pickler or to the customer.

Slabs are received from the caster and rolled from 9" thick down to selected thicknesses (<0.1" up to 0.5").

Slabs must first be heated prior to rolling. They are transferred onto the rolling table at 2300 F. They proceed through six 4-high roughing stands that reduce the thickness from 9" to 1 1/4" and produce a bar. The bar is then sent through the finishing mill (seven 4-high stands) which reduces the bar into a hot band. Exit speed is over 30 miles/hour. Cool water sprays onto the hot band to specific temperature ranges for coiling. Temperature at the last finishing stand and the coiler are closely monitored.

Pickler

Pickler

The Pickler supplies the Cold Mill or the final customer. Hot bands are run through the Pickler to remove rust and scale. The coils are run through baths of acid and then rinsed with water. The coils are oiled to prevent oxidation.

Cold Strip Mill

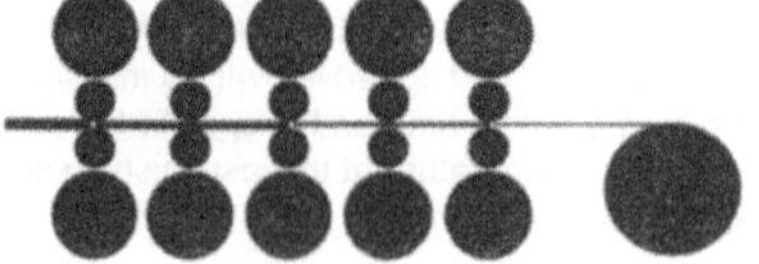

Cold Strip Mill

The Cold Mill supplies cold reduced coils to annealing, a continuous hot dipped coating line, or to the customer.

Hot Roll/Pickle coils are run through five 4-high stands which cold reduces the coil to a precise customer gage and also gives the coil a surface finish (pangborn).

Thickness is computer controlled and monitored.

Coil speeds are up to 60 miles/hour.

Box Annealing

Box Annealing

Box Annealing supplies fully annealed coils to the temper mill. The cold rolling process cold works the coil and increases its hardness. To improve the mechanical properties the coil must be heated in an annealing furnace. Coils are stacked on an annealing "base". A cover is placed over the coils and a non-oxidizing atmosphere (N_2 - H_2) is introduced.

A furnace is lowered over the cover. The furnace is fired to a prescribed cycle of time and temperature which is based on grade and size.

When the coils have cooled, the covers are stripped and the coils are sent to the Temper Mill.

Temper Mill

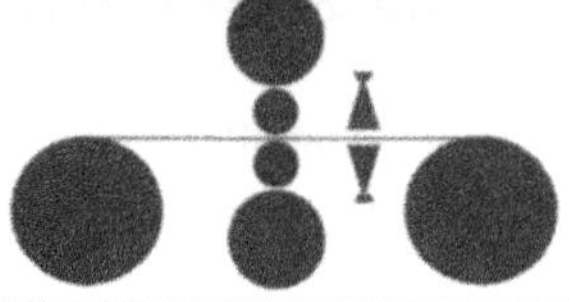

Temper Mill

The temper mill supplies finished coils to the customer, processor, Electrogalvanizing Line, or Terne Coat Line.

After annealing the coils are softened and may contain carbon and nitrogen in solution. This can lead to age hardening of the coil. A temper mill is used to retard the aging process and also to improve flatness and impart a uniform surface texture.

Coils are run through a 4-high stand which reduces the thickness of sheet approximately 1%. Rust preventative oil can be applied if requested.

Coils are sent to the customer or for further processing.

Porcelain Enameling Steels

- UNIVIT®
- I-F Enameling Steel
- VIT PLUS™ (Enameling Iron Replacement Steel)
- Cold Rolled Steels

Porcelain Enameling Steels

The following is a description of each product as it relates to AK Steel today. All enameling applications are considered to be Class I where the surface requirements today are nearly as severe as the automotive industry, especially with the newer porcelain enamel systems.

UNIVIT® - UNIVIT® is an aluminum killed, continuous cast, fully decarburized, flat rolled enameling steel developed by ARMCO (Pat No. 2,878,151, March 17, 1959). This steel is produced to be free from carbon boiling and enamel fishscale in a direct-on cover coat porcelain enamel system. Since this product was developed for the direct-on cover coat enamels which require a pickling operation, Cu and P are controlled to much tighter levels than on any other enameling steel when the end application requires a pickle type metal preparation system. This product is well suited for ground coat enameling systems and is especially good for the newer dry powder enamels because of its freedom from fishscale. I-F Enameling Steel has sag resistance better than UNIVIT®, and UNIVIT® has better sag resistance than VIT PLUS ™.

Porcelain Enameling Steels

I-F Enameling Steel - This is a vacuum decarburized, aluminum killed, titanium stabilized, continuous cast enameling steel. I-F Enameling Steel is suitable for ground coat enamels but will not develop adequate adherence in a direct-on cover coat application. It will be free from carbon boiling because the small amount of carbon left is tied up with the titanium. While this product is not guaranteed against enamel fishscale, recent research and subsequent chemistry and processing changes made recently at the mill have significantly reduced the susceptibility of this product to this enamel defect. It is our goal to guarantee the I-F Enameling Steel product against this defect in the near future. It has superior formability to UNIVIT® and to VIT PLUS™ products because of its inherently high r-value. Yield strength after firing porcelain is superior to UNIVIT® but not as good as VIT PLUS™. Sag resistance for this product is superior to both UNIVIT® and VIT PLUS™.

Porcelain Enameling Steels

VIT PLUS™ (Enameling Iron Replacement Steel) - Enameling Iron was a low metalloid, rimmed steel used exclusively in ground coat porcelain enamel applications. This product was only produced by AK Steel, Inland, and USS. It required the use of a strip normalizer to produce. Since all three domestic producers have shut these units down, this product is no longer produced today. (It has never been produced by offshore manufacturers). Since there is still a need for this type of product, castable replacement products have been developed. Our product is now commercially available and is called VIT PLUS™. This product is a controlled chemistry, aluminum killed, continuous cast steel which, at this time, is only produced through open coil annealing. (Future work will investigate production of this material through 100% hydrogen annealing facilities). As currently produced, VIT PLUS™ is suitable only for ground coat and two coat-one fire porcelain enamel applications. It is the only one of this type of product on the market today to be guaranteed against enamel fishscale. Of the three products discussed thus far, it is the worst for sag resistance. Yield strength retention is not as good as "the old" Enameling Iron product, but it is superior to UNIVIT® and, in some cases, to I-F Enameling Steel. Formability of this product is superior to the old Enameling Iron DQ, nearly as good as UNIVIT®, but not as good as I-F Enameling Steel.

Porcelain Enameling Steels

Cold Rolled Steel - Cold rolled steel is often used for non appearance application such as interior parts of appliances. These steels, historically, have most often been rimmed steels. No special processing or controls are used in the steel mill. Cold rolled steel is selected solely on price. The future of this type of product is tenuous because of the move in the steel industry to convert to 100% cast product lines. (Availability of rimmed cold rolled steels is nearly non existent today). As long as a rim zone is present, there is some protection from carbon boiling and a slight resistance to enamel fishscale. Cold rolled steels produced from cast products (which must be killed) do not have this protection. With rimmed steels, there is nonuniformity in properties from one end of the coil to the other due to both chemical and metallurgical variations from ingot top to bottom. There is also a greater tendency to warp and distort during the firing of the porcelain enamel with rimmed steels due to these variations. Due to the inherently metallurgically "cleaner" product and the typical low coiling temperatures used with today's aluminum killed cold rolled steels, there is a much greater likelihood of porcelain enamel fishscale in the final product. In general, AK Steel produces adequate product lines to satisfy the enameling industry and we do not recommend the use of cold rolled steel for porcelain enamel applications.

Lean Manufacturing Principles

Sean S. Reagan
Ferro Corporation, Cleveland, Ohio

Review and explanation of the principles behind lean manufacturing. This includes discussion on the creation of a value stream map and how this tool helps to identify waste in a process.

Introduction

What is lean manufacturing? Simply stated, lean manufacturing is the "elimination of waste" in any business process or "value stream." "Wherever there is a product for a customer, there is a value stream. The challenge lies in seeing it."[1] What do we consider to be waste? Waste can be defined as any activity in a business operation that does not add value to the product or service being delivered. A value-added operation is one that changes the shape or form of the product being delivered. If we look around us in our everyday world, there are numerous products among us. These products did not just appear, but some material or resource was utilized and transformed into the products we see around us.

The Eight Wastes

Through years of endless practice, eight wastes have been formally identified:[2]

- Waste of over-production is producing more than is needed. This results in more inventory, handling, space, interest charges, machinery, defects, overhead, people, and paperwork. Poor quality often is the root cause of over-production. LOOK for the quality hold area!
- Waste of waiting and delays is the need to wait for work, information, parts, tools, materials, equipment, etc. LOOK for groups of people in stand-up meetings.
- Waste of over-processing is the need to handle parts, tooling, and information more than once. This includes creating a product characteristic that may not be required by the end customer and inspecting work from a previous operation. LOOK for a parts-sorting area.
- Waste of motion is the movement of material, parts, or people that

does not add value to the product. Pick a process in your operation. LOOK to see how much motion is associated with moving a part or material through that process.

- Waste of transportation is the need to transport work or material in a process. LOOK for the number of required forklifts in a process.
- Waste of product defects is the need to rework products because of not making them right the first time. LOOK for the rework area.
- Waste of inventory is the need to store and hold more product than what is ordered. This includes work-in-process. The need to hold inventory is more than likely covering up other problems associated with the operation. LOOK in the warehouse and other storage areas to determine the level of inventory.
- Waste of people is not properly using the people associated with your operation. People are a company's most valuable resource. LOOK for people who are waiting or confused; hence, look for the questions.

Manufacturing Philosophies

If we consider the progress that industry and manufacturing has made over the past 80 years, we can begin to compare and contrast the differences between the philosophies employed then and now. These philosophies are outlined in Table I.

How Do We Become Lean?

So, how do we become Lean? Jim Womack has suggested that there are five crucial steps in becoming lean.[3] He has characterized these steps as "lean transformation:"

- Step 1: Find a change agent, someone who is able to see and understand the whole business process.
- Step 2: Find a guru and borrow a learning curve.
- Step 3: Seize (or create) a crisis to motivate action across your firm.
- **Step 4: Map the entire value stream for all of your product families!**
- Step 5: Pick something important and get started removing waste quickly, to surprise yourself with how much you can accomplish in a very short period.

Step 4 is highlighted because Womack points out that, although it is the most important, it is often the one that is overlooked. People are so anxious to begin eliminating waste that they do not have a clear picture of where they currently are and where they want to be in the future.

Table I. Lean Operation and Mass-Production Philosophy

Philosophy	Lean operation	Mass-production operation
Credo	"Manufacture the right product at the right time, in the right quantity, with the highest quality, at the lowest cost, and with the shortest delivery time."	"You can choose any color so long as it is black." (Henry Ford)
Quality	Typically very high by nature.	High volume will hide quality problems.
Changeovers	Very high.	Very low.
Product flow	Continuous, one piece flow; little wasted motion or transportation. PULL SYSTEMS EMPLOYED (supermarkets).	Noncontinuous, large batch flow; high degree of wasted motion and transformation. PUSH SYSTEM EMPLOYED.
Comfort level	Low, more difficult to react to unplanned, high-volume customer demand; however, will drive business to replace comfort with confidence.	High, easy to react to unplanned high-volume customer demand because of stockpiles of inventory; very comfortable.
Product line customization	Easier because of emphasis on product group flow centered around smaller run sizes on equipment tailored for quick changeovers.	Harder because of emphasis on process-oriented flow centered around large-batch processing equipment.
Lead time	Low, hours or days.	Long, weeks.
Days of inventory/ inventory turns	Days low, turns high.	Days high, turns low.
Accounting philosophy employed	Increased cash flow because of a commitment toward carrying lower inventories, which results in less interest paid on required debt to run business.	Cost/unit produced is lower because of large run sizes/fewer changeovers; greater utilization of resources.

Value-Stream Mapping

Therefore, if one of the most important steps in becoming lean is value-stream mapping, how do we do it? The first step is to create a current-state map. There are six general guidelines, or tips, to aid in this process:[4]

- Tip 1: Always collect current-state information while walking along the actual pathways of material and information flows yourself.
- Tip 2: Begin with a quick walk along the entire door-to-door value stream. One needs to get a sense of the flow and sequence of the process. After the quick walk through, go back and gather information at each process.
- Tip 3: Begin at the shipping end and work upstream, instead of starting at the receiving dock and walking downstream. This way you will begin with the processes that are linked most directly to the customer, which should set the pace for other processes further upstream.
- Tip 4: Bring your stopwatch and do not rely on standard times or information that you do not personally obtain. Numbers in a file rarely reflect current reality. File data may reflect times when everything was running well, for example, the first-time-this-year three-minute die change or the once-since-the-plant-opened week when no expediting was necessary. Your ability to envision a future state depends upon personally going to where the action is and understanding and timing what is happening. (Possible exceptions to this rule are data on machine uptime, scrap/rework rates, and changeover times.) One item to be cognizant of when using a stopwatch on the shop floor is to be aware of the sensitivity of the shop floor personnel. It may be better to simply watch and make a mental note of the time and record it. An argument to this may be that, in some processing systems, gains are being made at the tenths of a second. Fortunately, this is probably the exception, and, in this case, using a stopwatch is probably not practical nor does such an operation involve a human in most cases.
- Tip 5: Map the whole value stream yourself, even if several people are involved. Understanding the whole flow is what value-stream mapping is about. If different people map different segments, then no one will understand the whole. Remember, seeing the process as

a whole is key in eliminating waste. One cannot be focused on how an improvement benefits a particular department or operation.

- Tip 6: Always draw by hand in pencil. Begin your rough sketch on the shop floor as you conduct your current-state analysis, and clean it up later, again by hand and in pencil. Resist the temptation to use a computer.

Creating the Future-State Map

We now have created the current-state map. How do you take the next step toward creating the future-state map? In Part Two of *Learning to See,* Rother and Shook address how to create a lean value stream:[5]

- Look for areas of overproduction between process steps. Determine the takt time for the product family being mapped.

$$\text{Takt time} = \frac{\text{(available working time per day)}}{\text{(customer demand rate per day)}}$$

Are you producing at the takt time in your process?

- Develop continuous flow wherever possible.
- Try to use supermarkets (allows production control without scheduling/MRP) to control production where continuous flow does not extend upstream.
- Try to send the customer schedule to only one point in the production process. This point is called the "pacemaker process," because it sets the pace for the rest of the production process.
- Level the production mix (load leveling) at the pacemaker process. This will help ensure inventory availability as well as reduce lead time and improve your reaction time to changing customer demands.
- Create an "initial pull" by releasing and withdrawing small, consistent increments of work at the pacemaker process (level the production volume).
- Develop the ability to make "every part every x" in upstream processes of the pacemaker process.

Lean Tools

One thing that Suzaki points out is, "there is no cookbook approach" to being successful.[6] There is no one tool that is an instant provider of success. The most critical factor to success, however, is the involvement of all people, at all levels. Below are what could be considered to be tools one would want to have in their "lean toolbelt:"

- 5S, housekeeping: "A place for everything, and everything in its place."
- Developing standardized work: We need to track and measure how we are doing so that problems can be addressed when they arise.
- Statistical process control, six sigma, and continuous improvement.
- Jidoka, visual controls, automation, andon (warning lights), Poka–Yoke (mistake proofing).
- Total productive maintenance: Participation and involvement by all people (especially operators) in maintenance operations, i.e., cleaning and inspecting.
- Total quality control: Participation and involvement by all people in quality operations, for example, ISO and QS.

When we think of lean manufacturing, we should try to remember that it serves as an umbrella to encompass all aspects of an operation that contribute to eliminating waste.

References

1. Mike Rother and John Shook, *Learning to See*. The Lean Enterprise Institute, Mass., 2003.
2. Kiyoshi Suzaki, *The New Manufacturing Challenge*. The Free Press, New York, 1987.
3. James P. Womack and Daniel T. Jones, *Lean Thinking*. Simon and Schuster, USA, 1996.
4. See Ref. 1; p. 14.
5. See Ref. 1; pp. 41–54.
6. See Ref. 2.

ISO9001:2000 Process Mapping

Kara Joyce Kopplin
QTEC Consulting Corporation

One of the major changes from the 1994 edition of ISO 9000 to ISO9001:2000 is the emphasis on a process-based approach to quality and business management. What is the best way to map processes within your company? Fortunately, the standard allows quite a bit of latitude; therefore, companies have the freedom to design process maps that fit their unique needs. This paper presents the ISO9001:2000 process-mapping requirements and illustrates processes maps from several industries, along with development tools and auditing techniques.

A Familiar Philosophy

The new ISO9001 standard takes a practical approach to business management, pulling from the methodologies of several popular business management systems. The goals of this standard are continual improvement, defect prevention, and reduction of variation and waste, all of which are key elements in the Malcolm Baldridge Award, Six Sigma, and Lean Manufacturing.

Specifically, ISO9001:2000 is based on the plan, do, check, and act model. In fact, the four elements of the standard can be represented in this manner, where

- Resource management is the planning,
- Product realization is the doing,
- Measurement analysis and improvement is the checking, and
- Management responsibility is the acting.

The ongoing goal of this cycle is customer satisfaction!

A New Direction—The Process Approach

Fortunately, the new standard removes the compartmentalized, elemental requirements. The 2000 version recognizes that business is performed by conducting many processes, which have inputs and outputs from multiple activities, and typically involve several departments, people, tools, and pieces of equipment. This approach is first mentioned in the introduction to the standard (0.2 Process Approach),

"This International Standard promotes the adoption of a process

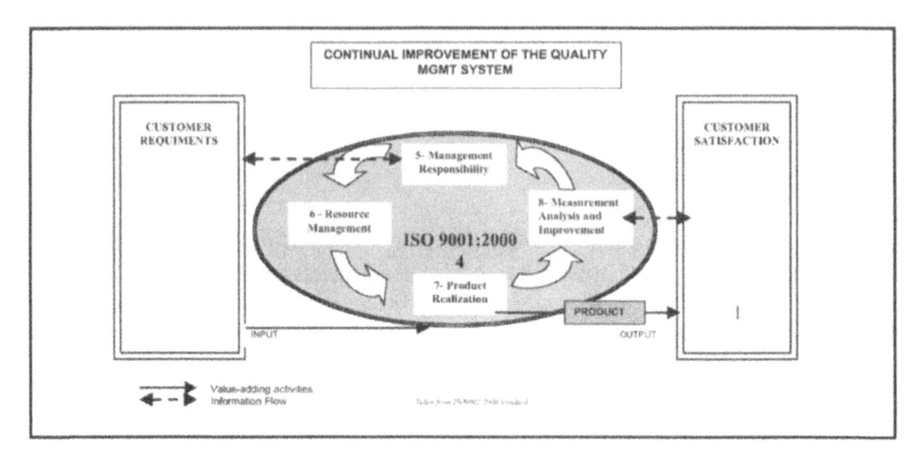

Figure 1. ISO9001:2000 continual improvement model.

approach when developing, implementing, and improving the **effectiveness** of a quality management system, to **enhance customer satisfaction** by meeting customer requirements." Reading further, we find the definition, "A PROCESS is a series of activities that transforms inputs into outputs."

Therefore, a single process may be represented this way.

The standard requires that these processes be

- Identified,

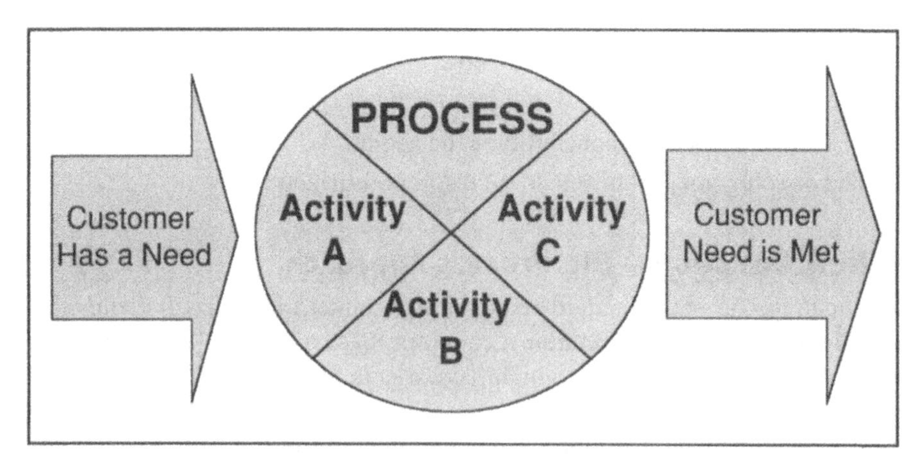

Figure 2. Example of a single process per the ISO definition.

- Mapped to show sequences and interactions,
- Measurable, with goals,
- Supported by resources and information,
- Monitored, measured, and analyzed for effectiveness, and
- Continually improving, because of actions taken in response to performance.

Create a Model of Customer-Oriented Processes

Because no further direction is given in the standard, companies are free to create their own process maps. A bird's eye view is the first place to start. Take a step back from the details and look at your business in broad terms. Focus on the MAIN processes in your facility, specifically the ones that have direct customer inputs and outputs and directly affect customer satisfaction. Consider in your organization, the

- Sales process,
- Product development process,
- Scheduling process,
- Manufacturing process,
- Shipping process, and
- Customer service process.

Keep in mind that these are not departments, but overall processes that involve several departments and specific activities. Your top-level map of customer-oriented processes (COPs) might look something like this:

By connecting the processes, the top-level map meets another requirement of the standard, that is, (4.1 General Requirements) the organization SHALL "… identify the processes needed, ... determine the sequence and interaction of these processes."

Add Support Processes

For more detail, you may include the processes that support your customer-oriented processes (support-oriented processes, or SOPs):

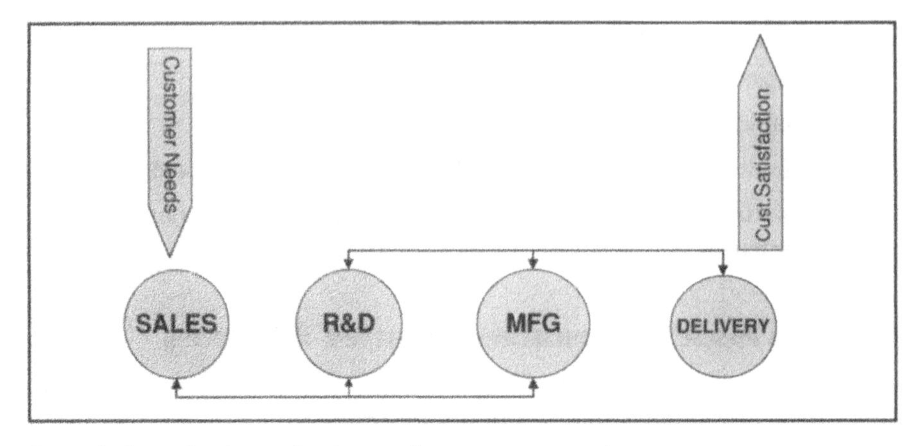

Figure 3. Example of a top-level map of customer-oriented processes.

Dive into the Customer-Oriented Processes

The next step is to dive into the details of your COPs. Consider enameling cast iron grates. If "manufacturing" is one of your customer-oriented processes, it may include several activities, such as

- Enamel preparation activities (batching , milling, testing, and setting up),
- Iron preparation activities (grinding, blasting, and annealing),
- Enameling (coating and drying/firing), and
- Inspection activities (dimensional and visual).

This can be captured using the basic process model illustrated earlier. If it is helpful to your organization, you can likewise map the support processes.

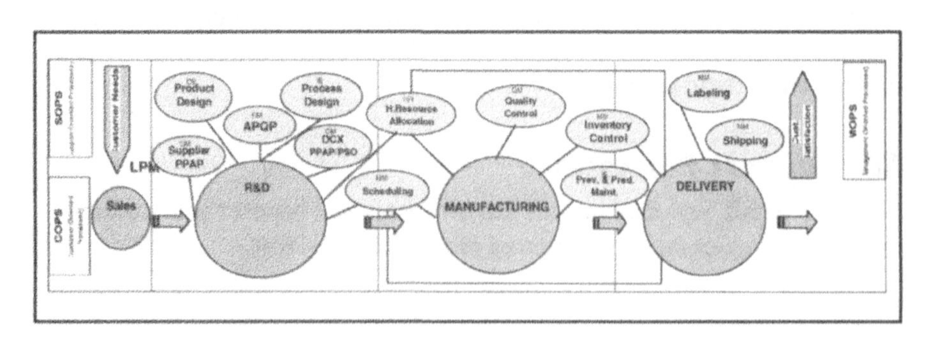

Figure 4. COPs and SOPs.

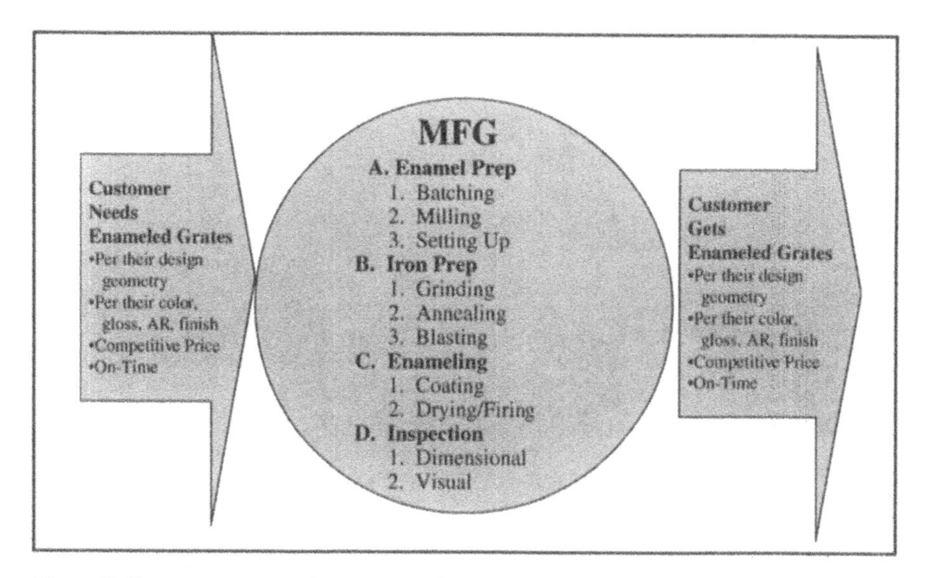

Figure 5. Customer-oriented process with inputs, outputs, and activities.

Dive Deeper and Explore All the Variables, with Help from Ishikawa

This tool, called the "turtle diagram," maps the inputs and outputs, as well as the variables that can affect process performance, much like the original fishbone diagram. The turtle identifies who does it, how it is done, and how it can be measured. (Use your imagination to see the turtle!)

Evaluate Process Performance

The new standard focuses on process effectiveness. To determine if a process is effective, it must be measured against a goal: (4.1 General Requirements) The organization. SHALL "... monitor, measure, and analyze these processes."

This is done by collecting and analyzing key measurable data, such as on-schedule product development, production scrap, and on-time deliveries. You might include these measurables directly in your maps:

The standard also requires the organization to "Implement actions to achieve planned results and continual improvement."

Management must review the key measurable data (often in a monthly

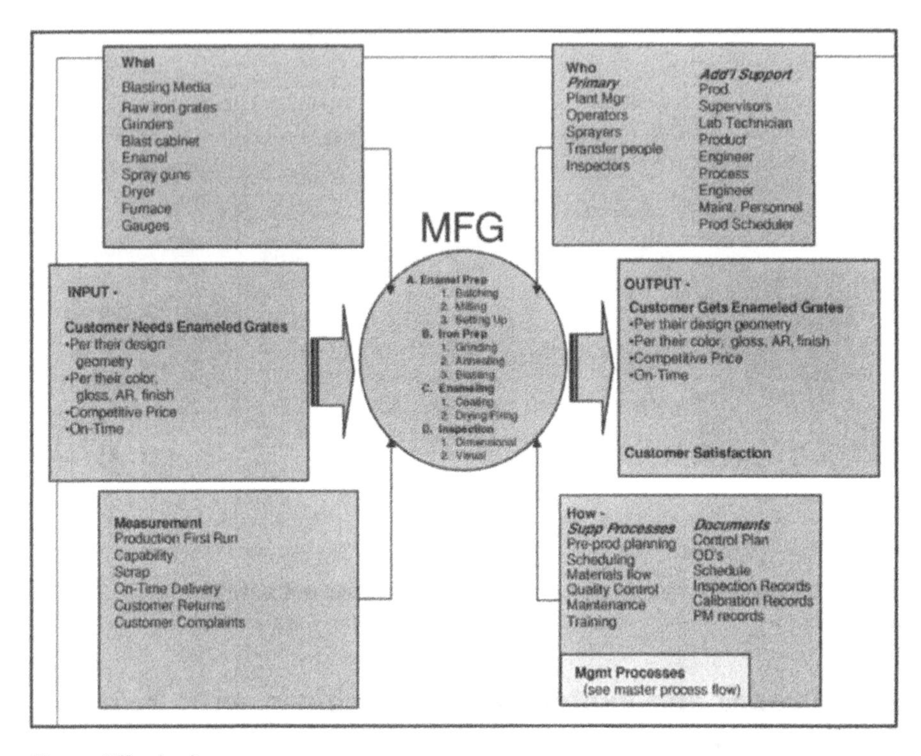

Figure 6. Turtle diagram.

review meeting). If measurables fall short of internal goals or the requirements of the customer, management must take action. This action can be corrective action, allocation of resources, training, etc. The standard does not require companies to be perfect, but it does require them to react when they do not meet goals! This is explained in the standard (4.1 General Requirements) that states the organization SHALL "... ensure operation and control of these processes are effective ..." and "... ensure availability of resources and information ... to support the operation and monitoring of these processes." Moreover, (5.6.3 Management) review output SHALL include decisions and actions related to "improvement of the effectiveness of the quality management system and its processes," "improvement of the product related to customer requirements," and "resource needs."

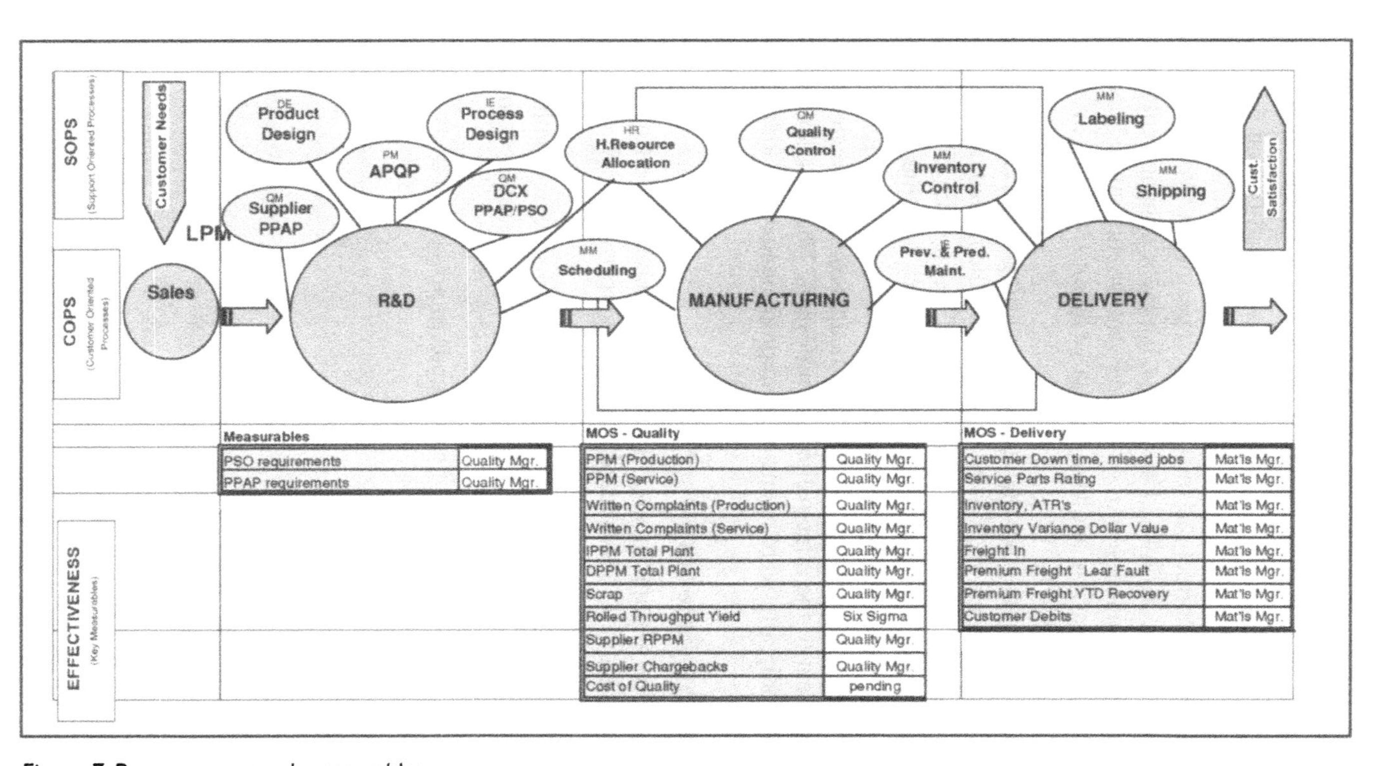

Figure 7. Process maps and measurables.

A New Way of Thinking—Process Auditing

Auditors must start by looking at the key measurables and asking a multitude of questions. Which ones did not meet goal (which processes are not effective)? Was the ISO/continual improvement/plan–do check model used? Do we correct our mistakes, or repeat them? Where in this cycle did the process break down? Do we have happy or unhappy customers? Are we delivering a nonconforming product? Are we missing delivery dates? Does management review and take corrective actions? Are the processes continually improving?

Such questions give the auditors a starting point in their investigations. The process maps and turtle diagrams give them road maps to follow to the root causes of the issues. In the course of the investigation, the auditors will follow where their questions and the answers lead them. In doing so, their paths will cross through multiple departments and functions in search of the true source of the breakdown. Documentation is still reviewed, but it takes a back seat to the effectiveness of processes. The standard or a "checklist" is used only afterward to ensure all requirements were addressed.

Creative Freedom

Below are multiple maps from different industries, all of which meet the requirements of the standard. Use some creativity, and create one that fits YOUR business!

This evolved from the initial process map several months after the certification audit. Maps are living documents and should change as the business changes.

Note the identification of continuous improvement tools under the manufacturing assembly process.

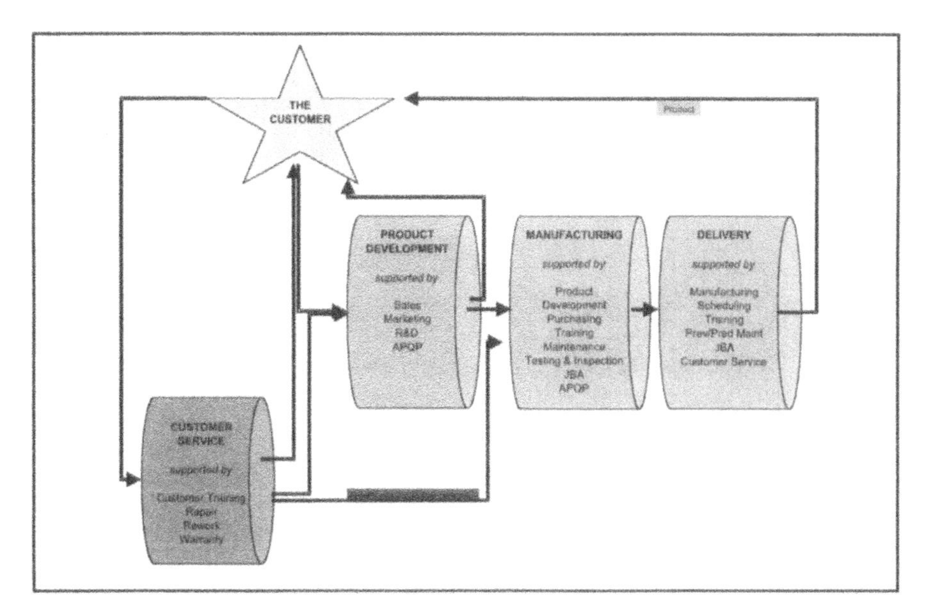

Figure 9. Medical device industry process map—take 1.

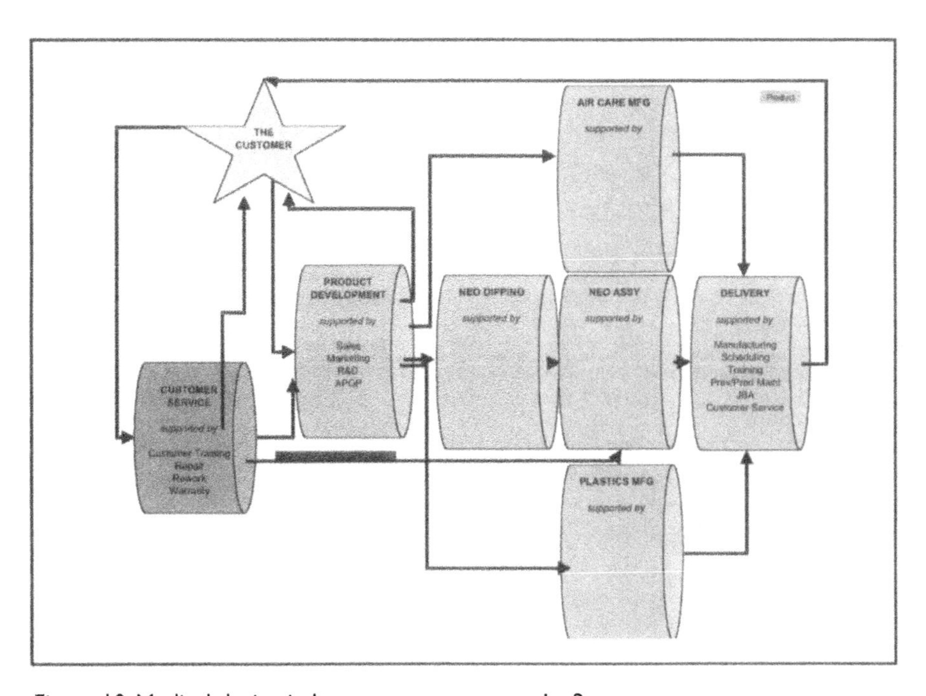

Figure 10. Medical device industry process map—take 2.

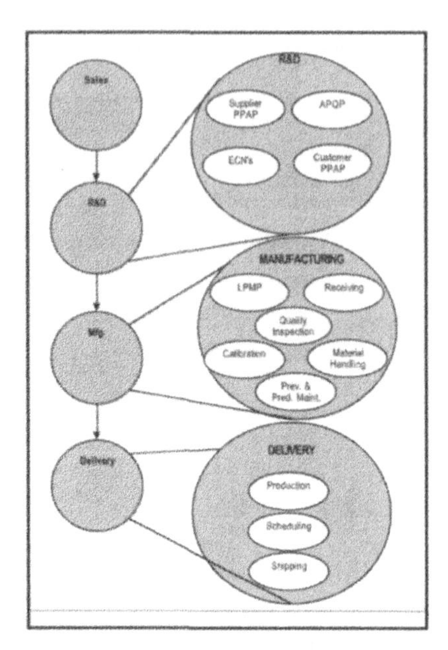

Figure 11. Microscope view process map.

	Customer Oriented Pro	Sales	R&D	Manufacturing	Delivery	Support Processes:	Customer Quotation	Contract Review	Planning	Purchasing	Mfg / Assy	Packaging	Design & Dev. Process	DFMEA	DVP&R	Supplier Development	Service Parts
Customer Oriented Processes																	
Sales			x	x			x	x	x	x		x					x
R&D		x		x					x	x		x	x	x	x		x
Manufacturing		x	x		x				x	x	x	x		x	x	x	x
Delivery				x					x	x	x	x				x	x
Support Processes:																	
Customer Quotation		x						x		x							x
Contract Review		x					x			x						x	x
Planning		x	x	x	x					x	x	x	x	x	x	x	x
Purchasing		x	x	x	x		x	x	x		x	x	x			x	x
Manufacturing / Assembly				x	x				x	x		x	x	x	x	x	x
Packaging		x	x	x	x					x	x		x			x	x
Design & Development Process			x						x	x	x	x		x	x	x	x
DFMEA			x	x					x	x	x		x		x	x	x
DVP&R			x	x					x	x	x		x	x		x	x
Supplier Development				x	x			x	x	x	x	x	x	x	x		x
Service Parts		x	x	x	x		x	x	x	x	x	x	x	x	x	x	

Figure 12. Automotive industry process matrix.

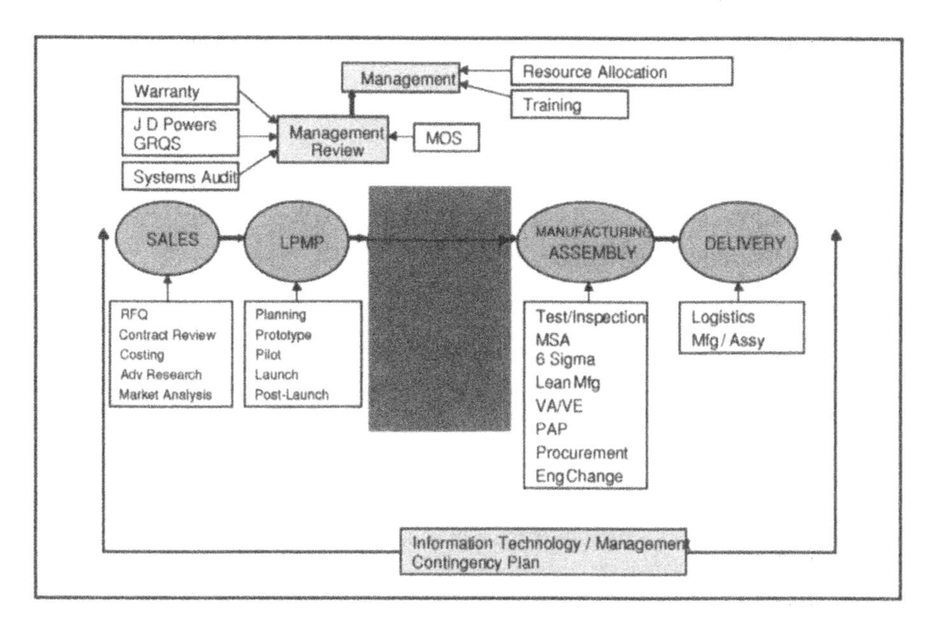

Figure 13. Automotive industry—assembly plant process map.

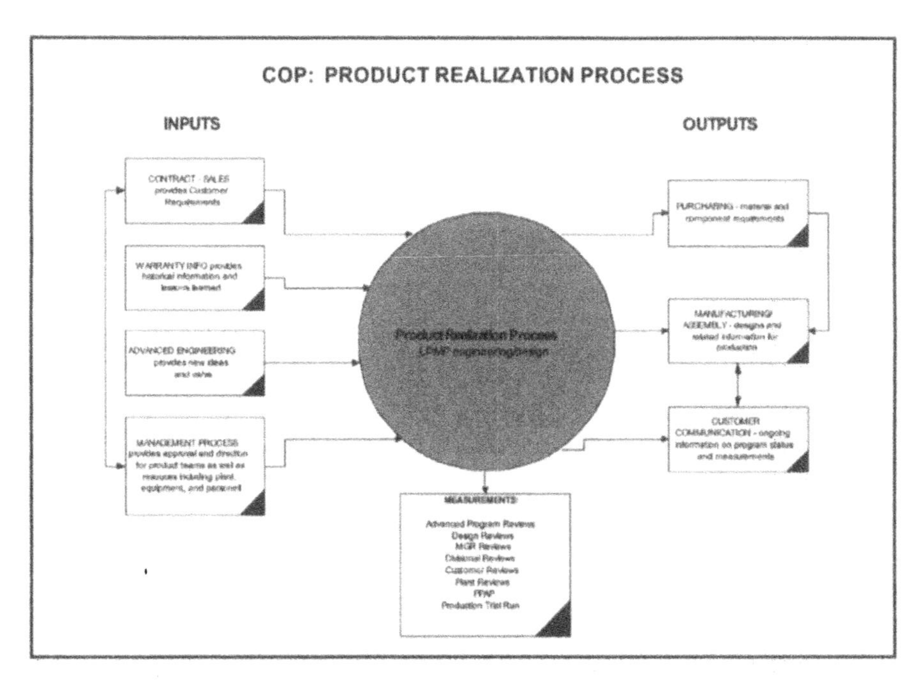

Figure 14. Automotive industry—engineering division COP map.

In Conclusion

Organizations have considerable freedom to create process maps that describe their systems, as long as they are

- Identified,
- Mapped to show sequences and interactions,
- Measurable, with goals,
- Supported by resources and information,
- Monitored, measured, and analyzed for effectiveness, and
- Continually improving, because of actions taken in response to performance.

By developing and using the process maps, your organization can easily monitor the effectiveness of its processes and identify weaknesses. Utilizing the methodologies of ISO9001:2000 results in

- Defect prevention,
- Reduction of variation and waste, and
- Continual improvement.

As a result, your processes will be more efficient, your customers will be happier, and your business will be stronger!

Your map does not have to be a visual illustration of your processes; it can be a matrix. This also meets the requirements of the standard.

Temperature Profiling: Problem Prevention, Not Just Problem Solving

Steve Sweeney
DATAPAQ, Inc., Wilmington, Massachusetts

In-process temperature profiling provides an accurate and comprehensive temperature profile of the complete product throughout the entire cure process. Temperature is one of the primary process variables in any heat-treating process. Temperature directly influences the structure and mechanical properties of the heat-affected substrates. Cure schedule confirmation, when performed postprocess, is done after the damage has been done and is tedious and time consuming. Rub tests, differential scanning calorimetry, impact tests, and gloss tests done at this time highlight problems but do not provide solutions. The benefits of in-process temperature profiling include quality assurance, process control, process optimization, improved productivity, improved efficiency, problem solving, and regulatory compliance.

Alternative methods of obtaining temperature data include pyrometric profiling and use of static and remote sensors. Pyrometric profiling is limited to substrates and not environmental temperatures, and static and remote sensors provide information only at discrete points in the process. No information regarding time at temperature is available.

An in-process temperature-profiling system is a complete hardware/software package comprised of a data logger, thermal barrier, thermocouples, and analysis software. Information for analysis includes maximum/minimum temperature, time at temperature, peak difference, slopes, process settings, graph options, customized reporting, and statistical process control. A profiling system can be used for direct and indirect profiling. Direct profiling analyzes an actual product during routine production, whereas indirect profiling involves the use of a test piece designed to imitate products in normal production.

Direct profiling is advantageous when it is too expensive or difficult to use a test piece, when products frequently vary in size, shape or coatings, when it is possible to apply thermocouples without damaging the product finish, or when sensors can be applied and removed quickly on-line during normal production.

Indirect profiling is suitable when thermocouples cannot be attached directly to the product without causing finish damage, when standard products are processed day after day, either when the line can be stopped or space provided for an addition of a test piece during standard process or when a test piece can be manufactured that accurately mimics a standard product.

Ideally, temperature profiling should be done routinely. The data obtained benefits the line operator and the QA/process manager. The line operator gains speed, ease of operation, automatic setup, data security, and simple and accurate initial QA check. The QA/process manager gets full comprehensive analysis, traceable reporting, historical review, and process control. Routine profiling helps spot problems before they impact production. Examples include determination of furnace conditions over time, adherence to specific rates of heating and cooling, identification of hot or cold spots, and analysis by zone.

In-process temperature-profiling systems can be configured for virtually every application. Durable and accurate data loggers are available with up to 10 selectable channels, 130,000 data readings, 0.1 seconds sampling rate, and an accuracy of ±0.5°F. State of the art two-phase thermal barriers, featuring 304 stainless steel, microporous insulation, and mullite textile lining, can withstand temperatures of 1835°F. The wizard-driven comprehensive analysis software is easy to use and fully customizable.

Specific features and benefits of the analysis software:

- "Reference Curve" shows the overall effect of any changes made to the furnace settings. It also shows any deterioration in furnace conditions over time.
- "Tolerance Curve" gives an instant visual picture of whether a process is out of tolerance.
- "View Temperature Data" instantly provides imbalances and hot and cold spots across the furnace width or around the load.
- "Calculate Slopes" ensures that there is adherence to the specific rate of heating or, more importantly, cooling.
- "Max/Min Temperature" gives an instant picture of each zone, of whether the process is within the specified limits.
- "Calculate Time at Temperature" provides a precise picture of the thermal balance of the furnace by showing which areas of the load, or sections of the furnace, "soak" at the temperature the longest.

Temperature profiling is proactive, not reactive. It provides crucial data needed to understand, control, validate, and optimize the thermal cure process. Profiling provides quality assurance of cure processes, optimization of line speed productivity, rapid fault finding, and validation of process control.

Topics

- Why is Profiling Important?
- Advantages of Profiling
- Alternative Methods
- Datapaq System
- Methods of temperature measurement
- Conclusion
- Questions & Answers

Why is Profiling Important?

Temperature Measurement

Temperature - primary process variables in any heat treating process.

It directly influences the structure and mechanical properties of the heat-affected substrates

Why is knowing the uniformity so important?

Critical Areas

Critical –

Component Temperature

not

Air or Oven Temperature

Cured or not?

Coating Properties

Physical

- **4 hardness**
- **4 flexibility**
- **4 impact resistance**
- **4 adherence**

Other

- **4texture**
- **4 color stability**
- **4 flow properties**
- **4 chemical resistance**
- **4 gloss retention**
- **4 service life**

Cure Schedule Problems

Under Cure

- ✖ Brittle
- ✖ Solvent attack
- ✖ Poor adhesion
- ✖ Too high gloss
- ✖ Gloss variability
- ✖ Poor Weathering

Over Cure

- ✖ Reduced Coating life
- ✖ Scorching
- ✖ Discoloration (yellowing)
- ✖ Too low gloss

Cure Schedule Confirmation Post Process

- 4 Experience / Guess Work
- 4 Rub Tests (Solvent Resistance)
- 4 Differential Scanning Calorimetry (DSC)
- 4 Impact Tests (Physical Attributes)
- 4 Gloss level (Control Panels)

Performed after the damage has been caused.

Highlights problems but with no solution.

Cure

Factors affecting the product heating rate:

- ➨ Oven Starting Temperature (Box Oven - Temperature drop on loading)
- ❊ Conveyor Track Speed
- ❊ Substrate thickness, weight or shape
- ❊ Mix of components / Oven Loading
- ❊ Oven Design (IR, Convection)
- ❊ Oven Temperature Recovery Rate
- ❊ Thermal Uniformity in Oven

Advantages of Profiling

In Process Temperature Profiling

Provides.......An Accurate and *Comprehensive temperature profile of the complete product throughout the entire cure process*

"In Process Temperature Profiling"

Benefits

- **Quality Assurance**
- **Process Control**
- **Process Optimisation**
- **Improved Productivity**
- **Improved Efficiency**
- **Problem Solving**
- **Regulatory**
- **Compliance**

Prevention .... Better than Cure

✓ Routine Profiling
....Once Day / Shift

✓ Reproducible Measurement
....Test Piece Use

✓ Quick Efficient QA

⇩

Characterization of Oven / Process Control Status

Alternatives Methods of Profiling

Topics

- Why is Profiling Important?
- Advantages of Profiling
- Alternative Methods
- Datapaq System
- Methods of temperature measurement
- Conclusion
- Questions & Answers

Static and Remote Sensors

The Limitations:

✖ **Inaccurate Temperature Measurement:**

Product temperature estimated from oven temperature therefore no consideration of thermal mass etc.

✖ **Incomplete Temperature History**

Provides temperature information at discrete points in the process. No information regarding time at temperature.

Datapaq System

Data Logger

- **Highlights**
- Measurement range 32 to 2500°F
- ~60,000 data points
- 0.1 sec to 10 minute
- NiMH battery:
- Start/Stop buttons allow greater user control
- Line of sight LED lights to easily check data logger status

Thermal Protection

Operating temperature up to 1560°F

- 304 stainless steel
- Microporous insulation
- Mullite textile lining
- 23 lbs
- 1475°F@ 1.00 hr

Thermocouples

- Max Temperature: 2280°F
- Thermocouples are (ANSI MC 96.1) Special Limits of Error

Information Analysis

Max / Min Temperature

Datapaq Value (Index of cure)

Time at Temperature

Process Settings

View Temperature Data

Graph Options
(3D, Reference, Tolerance)

Peak Difference

Customized Reporting

Slopes

Statistical Process Control

Methods of Temperature Measurement

Profiling Methodology

What to measure ?

❶ **Direct Profiling**
Profile of Coated Product during Routine Production *(Product Destined for Customer)*

❷ **Indirect Profiling**
Profiling of a Test Piece designed to imitate products in normal production
(Used repeatedly for profiling)

Direct Profiling

➧ ***Too expensive or difficult to use a Test Piece (large products ie. Automotive)***

➧ ***Coated products vary in size / shape or coating frequently***

➧***Possible to apply thermocouples without damaging a products finish (Automotive - internal body shell)***

➧ ***Repeatability of Measurement***
Need sensors that can be applied and removed quickly on-line during normal production (Magnetic or Clip Probes)

In-Direct Profiling

- ➧ *Thermocouples cannot be attached directly to the product without causing coating damage*
- ➧ *Standard product processed day after day*
- ➧ *Line can either be stopped or space provided for addition of test piece during standard process.*
- ➧ *Test piece can be manufactured that mimics accurately a standard product*
- ➧ *Reproducible Test Conditions Guaranteed*
- ➧ *Easy Quick Set-up*

Test Piece - Design / Use

- ➧ **Material** *(Thermal Conductivity)*
- ➧ **Thermal Mass** *(Size and mass match)*
- ➧ **Surface Coating ?** *(Essential for IR)*
- ➧ **Geometry**
- ➧ **Orientation**
  (Match conductive heat transfer characteristics)
- ➧ **Oven Loading**
  (Test piece profiled during standard production run)

Conclusion

Temperature Profiling

......generating the information needed to understand, control, validate, and optimize your thermal cure processes

- ◎ **Quality Assurance of cure process**
- ◎ **Optimisation of line speed productivity**
- ◎ **Rapid Fault Finding**
- ◎ **Validate Process Control (ISO Docs)**

Proactive not Reactive

Quality Systems for Cast Iron Enameling

Robert Hayes
Porcelain Industries

Successful coating of a cast iron substrate with porcelain enamel requires an understanding of every step in the process. This begins with the chemistry of the molten iron and concludes with the acceptance of the final product by the customer. The steps in between are numerous, and there are many critical process variables (CPVs) that must be controlled.

Does your company have a formal, documented quality system? If so, is it effective?

Quality systems exist to help you control your processes. How is this accomplished? Let us look at some of the elements of a quality system:

- Control of nonconforming product;
- Material ID and status;
- Drawing/specification control;
- Corrective action;
- Statistical methods;
- Employee training;
- Calibration and gage control;
- Supplier quality audit;
- Internal quality audits;
- Deviation authorization; and
- Receiving inspection.

This list is not meant to be all-inclusive, just some of the main elements.

Take "control of nonconforming product" for instance. What do these parts cost you if they get sent to your customer? Will you be subjected to product liability problems? Is it possible that serious injury or death may occur because you allowed defective product to be shipped?

What about "corrective action"? How many times has your response been to retrain the operator? Does that get the job done? You must strive to find and implement a root-cause solution that actually *prevents* the problem from recurring. Simply discussing the situation with the operator rarely accomplishes that goal.

How well do you train your people? If you are the new guy, what training will you receive before being placed in the job? Do you rely on the new persons' co-workers to train them? If so, then what do you suppose they are being told? Do you have documentation for the job? Does the new hire get to see it? Does the new hire understand it? Do new persons demonstrate a minimum level of competence before being released to work on their own?

Audits—do you audit your suppliers? Do you give them specifications, and are they holding them? How do you know if you do not audit their results? Do you audit yourself? Many times companies have very good documentation but fail to follow through by using it. You do not know until you perform internal audits to your procedures.

Are you audited by your customers? What about external standards, such as ISO and QS? A poor score on a customer audit can jeopardize your relationship and the amount of business you may get in the future.

Your quality system must have buy-in from all levels of employees. Certainly, if the production line people do not accept it, then your processes will drift out of control. But what about management? If the top-level managers do not "sing the praises," then mid-level managers, supervisors, and so-on down the line will not use the discipline necessary to make the system work. Your people must believe in it in order to make it work because it is an everyday effort.

How do you control your processes? Everyone knows that this is something that must be done, but what steps must be taken to accomplish it? Design of experiments and statistical process control are good tools when implemented properly. Be aware that simply having operators making charts does not improve the process. You have to demonstrate to the floor-level employees how this tool will help them, or it will not work.

There is no substitute for knowing the process. What are your CPVs? Are you holding your own specifications? What is your Cpk on the critical CPVs of a given process? You need to know the answers to these questions if you are to control the process.

Document each process. If you have employee turnover (who does not?), how will you pass along all the various settings and methods used? If your process is not documented, then your ability to produce a good product at an acceptable yield walks out the door with the employee who just quit!

An example of how we document any given part is shown here.

Why is all this important?

Usually your customers' specifications will require that you have a good quality system in place. If the process is not documented, if employees are not properly trained, and if someone does not go around double checking, then the product you produce likely will have an excessive amount of variation and likely will exceed those specifications.

Having acceptable scores on customers' quality audits opens the door for more business. Once others in the industry hear of your success, more opportunities will follow.

Having a good, functioning quality system will reduce your scrap and rework costs and, therefore, improve your company's bottom line.

Porcelain Industries	Cast Iron Process Instruction	
Casting P/N:	Date:	Rev.:
Quality:	Production:	Anneal?

RAW				
Machine	Seconds	Pcs./Rod	Rotate 90 deg?	Instructions
Blastec				
Goff				
Tumble				
REBLAST				
Blastec				
Goff				
Tumble				

Application Data		
Enamel:	Spec. Gravity:	Slump:

Gun #	1	2	3	4	5	6
X						
Y						
Z						
R1(Deg)						
R2(Deg)						
PSI(F/A)						
Fan						

Chain Speed Hz:	Left Spindle Hz:	Right Spindle Hz:	Tank PSI:

Dry Thickness Targets (mils):

Transfer/Firing Data:

Inspection Instructions:

Packing Instructions:

Picture of Part:	Individual Part Numbers: Gloss Black: Gloss Grey: Set Part Numbers: Gloss Black: Gloss Grey:	Rev. A: Release

Form Rev. F   ref. proc. # 006

Effects of Spray Patterns on Powder-Coating Thickness

Ralph Gwaltney
Maytag Cleveland Cooking Products

Abstract

High-output automatic powder-spraying systems are the norm for major appliance producers today. As the speed and volume of production processes increase, there is a greater need to effectively use the powder-spraying equipment to achieve complete parts coverage. A spreadsheet program has been developed to address the need to optimize the gun-spraying patterns on a variety of appliance parts.

Variables that affect the gun-spraying patterns are number of guns, conveyer speed, gun spacing, part height, spray width, part shape, nozzle shape, part design, gun motion, and powder.

Gun set-up formulas have been developed to determine the number of fixed guns required, gun distance from part, spacing between guns, number of moving guns required, gun mover stroke, and gun mover speed.

Examples of differing spray patterns are illustrated in the accompanying diagrams. The effect of gun mover speed and pattern overlap is shown as well as the effect of multiple guns. Thickness on test parts is illustrated for application at 0, 12, 21, and 24 oscillator strokes/min. Higher oscillator speeds produced more-uniform application patterns. In contrast, systems using reciprocating machinery to move guns exhibit sharper patterns (at travel direction change) and possibly a greater degree of coverage mismatch compared with a smoother pattern developed by oscillators.

Specific Gravity

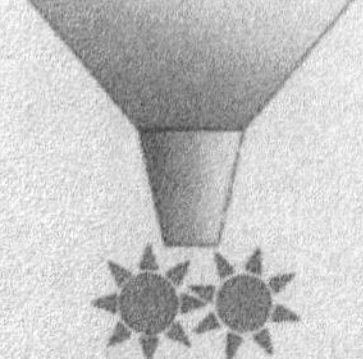

Powder Coating Thickness
Acceptable Production

Finishing Variables

Gun Set-up

Powder Coating Thickness

Powder Spray Gun Set-up Variables

- Number of guns
- Gun spacing
- Spray width
- Nozzle shape
- Gun motion
- Conveyor speed
- Part height
- Part shape
- Part design
- Powder

Gun Set-up Formulas

There are formulas to determine…

- Number of fixed guns required
- Gun distance from parts
- Spacing between guns
- Number of moving guns required
- Gun mover stroke
- Gun mover speed

Gun Movers

- Oscillator

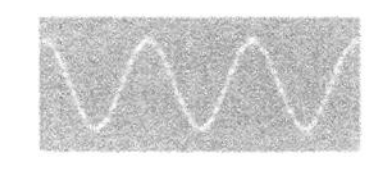

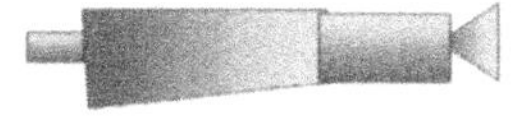

- Reciprocator

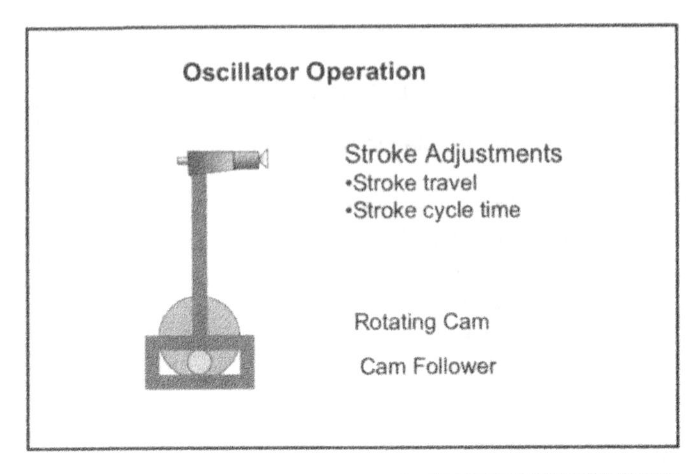
Oscillator Operation
Stroke Adjustments
•Stroke travel
•Stroke cycle time
Rotating Cam
Cam Follower

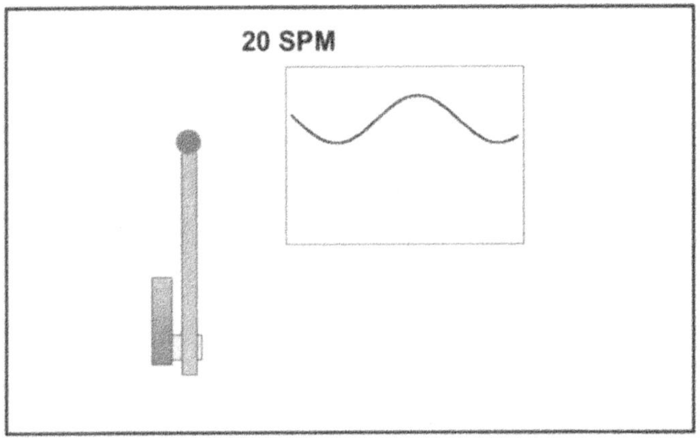
20 SPM

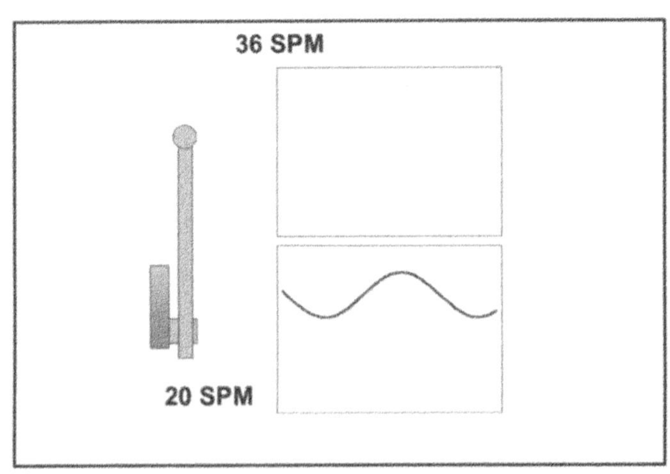
36 SPM
20 SPM

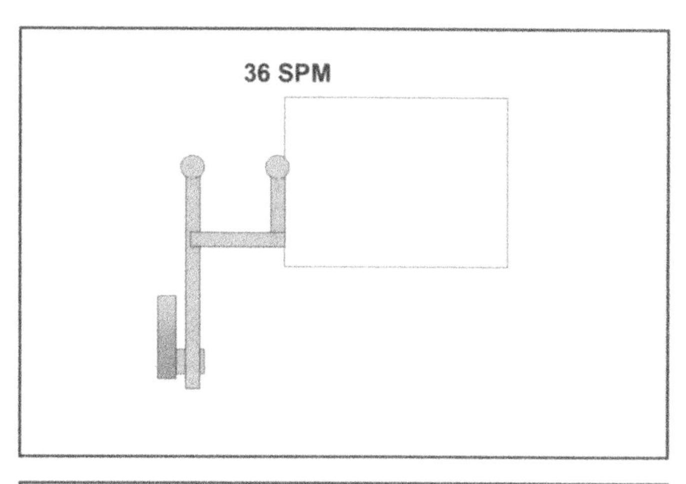

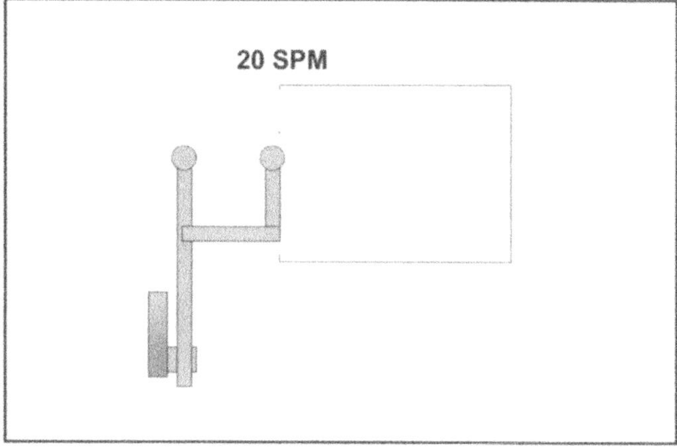

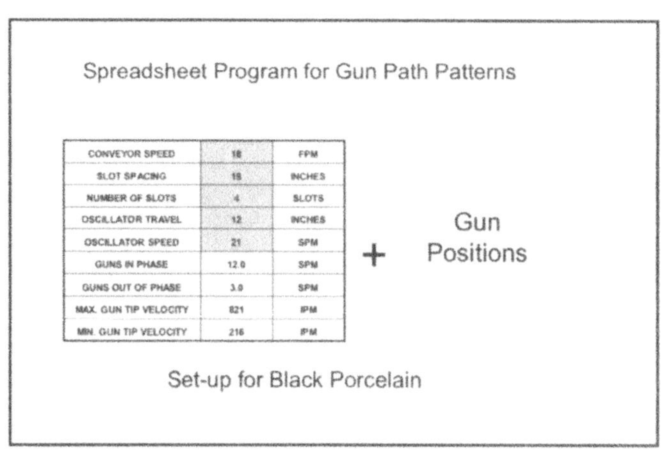

CONVEYOR SPEED	18	FPM
SLOT SPACING	18	INCHES
NUMBER OF SLOTS	4	SLOTS
OSCILLATOR TRAVEL	12	INCHES
OSCILLATOR SPEED	21	SPM
GUNS IN PHASE	12.0	SPM
GUNS OUT OF PHASE	3.0	SPM
MAX. GUN TIP VELOCITY	821	IPM
MIN. GUN TIP VELOCITY	216	IPM

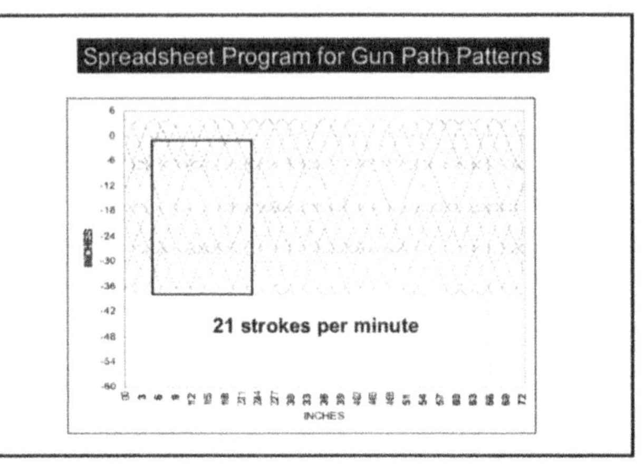

Spreadsheet Program for Gun Path Patterns
21 strokes per minute
INCHES
INCHES

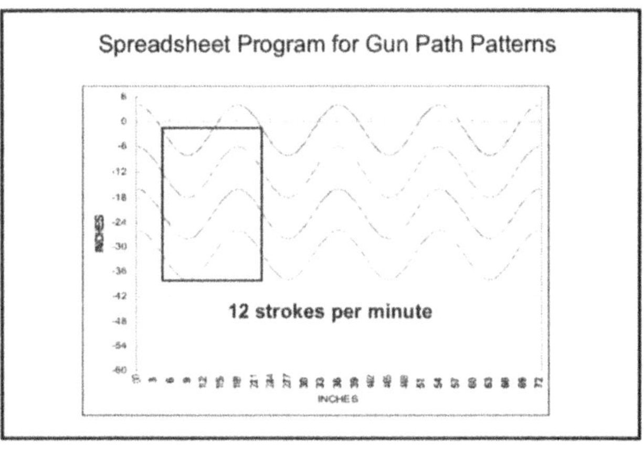

Spreadsheet Program for Gun Path Patterns
12 strokes per minute
INCHES
INCHES

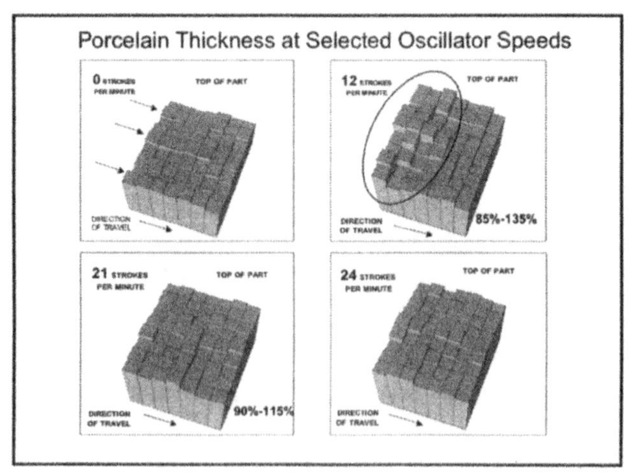

Porcelain Thickness at Selected Oscillator Speeds
0 STROKES PER MINUTE
TOP OF PART
DIRECTION OF TRAVEL
12 STROKES PER MINUTE
TOP OF PART
DIRECTION OF TRAVEL
85%-135%
21 STROKES PER MINUTE
TOP OF PART
DIRECTION OF TRAVEL
90%-115%
24 STROKES PER MINUTE
TOP OF PART
DIRECTION OF TRAVEL

Cost Comparison of Porcelain Enamel Powder versus Organic Powder Paint

Jeffrey Sellins
Maytag Cleveland Cooking Products

Mike Horton
KMI Systems, Inc.

Abstract

Porcelain enamels are often perceived as being a premium-priced coating system because of the high temperature of processing and the long-term durability. This study was undertaken to compare the cost of enameling and painting similar articles. Metal preparation (degreasing), rinsing and drying, curing or firing, process maintenance (hanger cleaning), and personnel needed to manage the systems were included.

Recent changes and improvements in the enamel operations at Maytag Cleveland Cooking Products, such as "no-transfer" operations that eliminate the hand movement of parts from the application line to the furnace line and more-efficient processing equipment, have made enameling a good choice for appliance and cooking products producers. The no-transfer operation allows the powdered enamel to be fired on the application line hooks. The hooks are subsequently sandblasted or thermally shocked to remove the fired-on enamel when an excess of enamel is accumulated on the hooks. This would be analogous to the pyrolysis of organic coating hanging hooks to remove accumulated paint.

A summary of the cost components (energy for cleaning and curing or firing, labor, and scrap costs) indicates that a no-transfer enameling system was within 1 cent/ft^2 of the cost of a paint coating for energy demand. A typical parts transfer system for porcelain enamel was about 6 cents/ft^2 higher than a paint system, regarding energy. Regarding a range of paint types (low cost, general purpose, and high temperature) curing at 500, 750, and 1000°F, no differences in total costs were shown.

Overall costs for porcelain enameling with a no-transfer system was \$0.0234 more than the lowest-cost paint system and \$0.0198 more than the highest-value paint system. The typical transfer system was \$0.0346 more expensive than the lowest-cost paint system and \$0.031 more than the highest-value paint system.

Noneconomic considerations, such as perceived value, durability, and product characteristics, including heat resistance, scratch resistance, and color stability, are important and need to be considered when choosing a finish. Porcelain enamels excel in these attributes.

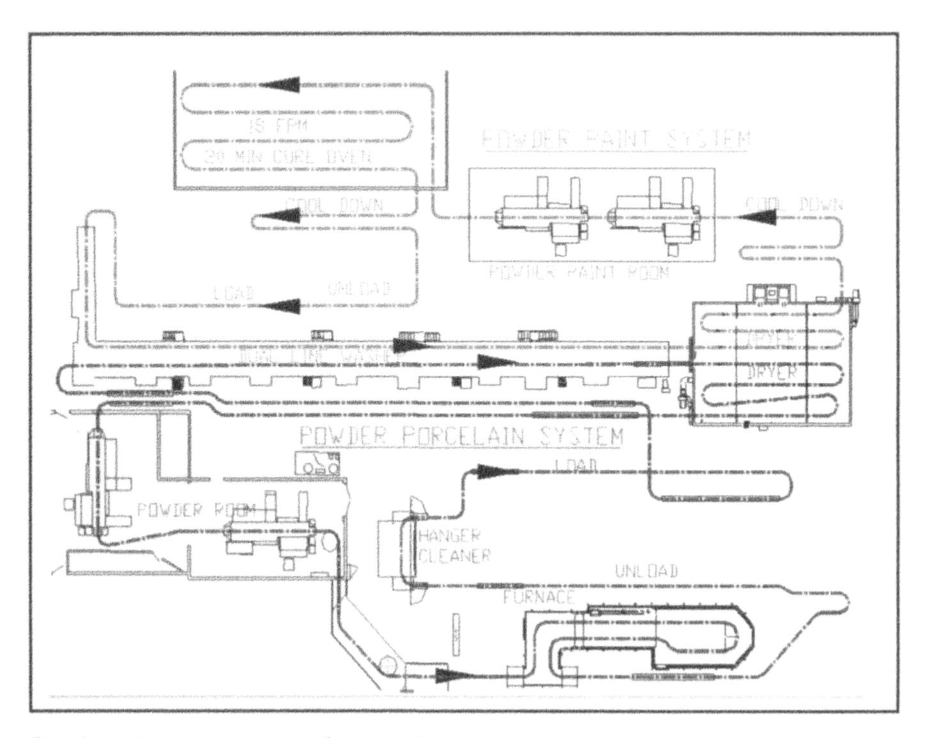

Powder paint system vs. powder porcelain system.

SYSTEM DESIGN REQUIREMENTS FOR COMPARISON

DESIGN SPEEDS: 15 TO 18 FEET PER MIN.
PART SIZE: 24" HIGH X 30" LONG
PARTS/HANGER: 4 PARTS
HANGER CENTERS: 36 INCHES
HANGERS PER HOUR: 360
PARTS PER HOUR: 1,440
PART PACKAGE: 54" HIGH X30" LONG
COATING FT. SQ. PER HOOK: 20
COATING FT. SQ. PER MIN.: 120
PART WEIGHT PER HANGER: 30
PART POUNDS PER HOUR: 10,800
HANGER POUNDS PER HOUR: 5,400
CONVEYOR POUNDS PER HOUR: 7,200

PROCESS DESCRIPTIONS COATING SYSTEM COMPARISON

PROCESS	POWDER PORCELAIN		POWDER PAINT COATING RATING °F
	TRANSFER	NO-TRANSFER	400,750,1000 °F
1. LINE SPEED	15 FPM	18 FPM	25 FPM
2. PART LOADING	4 OPERATORS		4 OPERATORS

PROCESS DESCRIPTIONS COATING SYSTEM COMPARISON

	POWDER PORCELAIN		POWDER PAINT	
3. PRETREATMENT	TIME(SEC)	TEMP.	TIME (SEC)	TEMP.
PRECLEAN	15	AMB.	55	AMBIENT
CLEAN	60	150°F	52	150°F
CLEAN	30	150°F	0	AMBIENT
RINSE	45	140°F	65	140°F
IRON PHOSPHATE	0	AMB.	63	140°F
RINSE	30	AMB.	76	AMBIENT
SEALER	0	AMB.	33	120°F
DI HALO RINSE	0	AMB.	5.5	AMBIENT
SLUDGE REMOVAL	NONE		IRON PHOS. FILTER	

PROCESS DESCRIPTIONS
COATING SYSTEM COMPARISON

	POWDER PORCELAIN	POWDER PAINT
4. DRY-OFF OVEN		
TEMPERATURE	400 °F	400 °F
TIME	5 MIN	5 MIN
5. AMBIENT COOLING	8 MIN	8 MIN
6. POWDER COATING		
AUTOMATIC GUNS	30	8
MANUAL GUNS	2	2

PROCESS DESCRIPTIONS
COATING SYSTEM COMPARISON

	POWDER PORCELAIN		POWDER PAINT
7. PART TRANSFER			
OPERATORS	5	NONE	NONE
8. FIRING/CURING			
TIME(MIN)	2.33	1.95	20
TEMPERATURE (F)	1560°F	1560°F	400 TO 500°F
9.AMBIENT COOL DOWN TIME (MIN)	8		8

PROCESS DESCRIPTIONS
COATING SYSTEM COMPARISON

	POWDER PORCELAIN		POWDER PAINT
10. PART UNLOADING OPERATORS	3		3
11. HANGER CLEANING	ON LINE BANGER	ON LINE SHOT	OFF LINE BURN OFF OVEN
	8 MIN		8 MIN
12. HANGER CLEANER OPERATORS	NONE	1	2 ON OFF SHIFT

COATING APPLICATIONS AND COMPARISONS					
	POWDER PORCELAIN		POWDER PAINT		
	TRANSFER	NO-TANSFER	400	750	1,000
APPLICATION (MILS)	4.0	4.0	1.75	1.75	1.75
COVERAGE (FT^2/MIL)	70	70	100	100	100
MATERIAL COST/LB	1.40	1.40	$2.50	$6.00	$8.00
FIRST PASS YIELD	92%	96%	98%	98%	98%
COATING MAT. COST/FT^2	$0.087	$0.083	$.045	$0.105	$0.14

COATING APPLICATIONS AND COMPARISONS					
	POWDER PORCELAIN		POWDER PAINT		
	TRANSFER	NO-TANSFER	400	750	1,000
PRETREATMENT MATERIAL COST	$0.0005	$0.0005	$.002	$.002	$.002
LABOR COST/SQFT	$0.075	$0.06	$0.06	$0.06	$0.06
MATERIAL COST/LB	$1.40	$1.40	$2.50	$6.00	$8.00
STEEL COST/SQFT	$0.348	$0.348	$0.33	$0.33	$0.33
HANGER CLEANING COST	$0	$0.007	$.004	$0.004	$0.006
SCRAP COST/SQFT	$0.004	$0.002	$.001	$.001	$.002

ENERGY COST COMPARISONS					
GAS= $8.5/1000 FT_	POWDER PORCELAIN		POWDER PAINT		
ELECTRIC= $0.05/KW-HR	TRANSFER	NO-TANSFER	400°F	750°F	1,000°F
PRETREATMENT GAS COST/SQFT	$0.0043	$0.0043	$.0058	$.0058	$.0058
DRY-OFF OVEN COST/SQFT	$0.0027	$0.0027	$.0027	$.0027	$.0027
CURE OVEN GAS COST/SQFT	$0	$0	$.0029	$.0035	$.0035
FURNACE GAS COST/SQFT (PREHEAT AIR)	$0.0084	$0.0067	$0	$0	$0
TOTAL ELECTRIC COST/SQFT	$0.0013	$0.0018	$.0012	$.0012	$.0012

TOTAL COST COMPARISON					
	POWDER PORCELAIN		POWDER PAINT		
	TRANSFER	NO-TANSFER	400°F	750°F	1,000°F
TOTAL COST/SQFT	$0.4442	$0.433	$.4096	$.4102	$.4132

COATING PROPERTY CHARACTERISTICS		
PROPERTY	POWDER PORCELAIN	POWDER PAINT
ADHESION	VERY GOOD	GOOD
CORROSION	SUPERIOR	500 HOUR SALT SPRAY
HARDNESS	SUPERIOR	4 H PENCIL OR LOWER
HEAT RESISTANCE	1000°F PLUS	400 TO 1000°F
GLOSS	HIGH GLOSS	400°F HIGH GLOSS 750°F LOW GLOSS
DESIGN RESTRICTION	SOME REQUIRED	LESS RESTRICTIONS

COATING PROPERTY CHARACTERISTICS		
PROPERTY	POWDER PORCELAIN	POWDER PAINT
PROCESS CONTROL	NO TRANSFER SIMILAR TO POWDER PAINT & REQUIRES MORE POWDER GUNS	PRETREATMENT REQUIRES MORE CONTROLS
ENVIRN. CONCERN	MODERATE TO NONE	PRETREATMENT OF CONCERN
FUTURE ECONOMIC CONCERNS	NATURAL GAS & RAW MATERIALS	OIL PRICES AND RAW MATERIALS
FUTURE MARKET TRENDS	HIGHER QUALITY, HIGHER MARKET VALUE, PE QUALITY MARKET	LOWER COST = ?

Reducing Fishscale in Hot Water Tank Enamels

Mike Wilczynski and Roger Wallace
A.O. Smith Corporation

Fishscale is delayed chipping (in a shape resembling a fishscale) of the fired and cooled enamel finish. During firing, hydrogen enters the steel in the atomic form. Hydrogen is introduced into the steel from various sources, such as hydrogen in the form of moisture in the furnace atmosphere, improperly dried bisque, water in clays or mill additions, residue left during metal preparation, and products of combustion. During firing, this moisture reacts with the steel in the following manner:

$$H_2O + Fe \Rightarrow 2H^+ + FeO$$

As the temperature increases, the solubility of atomic hydrogen (H) in steel increases to as much as 1000 times that at room temperature. As the enameled part cools, the solubility of hydrogen decreases. The hydrogen that was dissolved at the higher temperature now begins to diffuse and form molecular hydrogen (H_2) at the steel/enamel interface. If the pressure of the hydrogen gas builds up enough to actually overcome the enamel adherence or the strength of the enamel itself, then fishscale will occur.

There are several factors that affect the occurrence of fishscale in an enamel application that involve sources of hydrogen. One factor is the level of adherence of the enamel to the steel. The greater is the adherence, the greater is the resistance to fishscale. Another factor is the number of bubbles in the enamel layer. Fewer bubbles generally mean more fishscale. The frit and the types of clay used determine the number of bubbles. A third factor is the enamel's ability to chemically tie up the atomic hydrogen before it forms a gas as molecular hydrogen. This can be accomplished through the use of certain mill-added chemicals. The type of steel also can be a factor. Some steel can trap hydrogen molecules in its internal voids, thus preventing fishscale.

Recently, we needed to come up with a solution to a fishscale problem at a water heater manufacturer. The manufacturer of water heaters was experiencing major fishscale problems on all its fired ware, including tank shells and gas flues. Several proposals to reduce the amount of fishscale, such as increasing enamel adherence with a different frit and changing the enamel clays to increase bubble size, were discussed. We suggested using a proprietary, nickel-based material (designated as N-133) that has the ability to reduce, if not completely eliminate, fishscale. The N-133 works as a catalyst to oxidize residual hydrogen gas at the steel/enamel interface to water (H_2O).

First, we needed to produce evidence that these changes would reduce fishscale before we proceeded to a production trial. A test was developed that could quantify an enamel's or a steel's ability to resist fishscale. The test is very severe and ultimately produces fishscale in even the most resistant enamels or steels. The test involves exposing the bare-steel side of a test panel that has enamel on the other side to an acid solution. An electrolytic cell is used to generate large amounts of hydrogen gas from the reaction of the acid and the steel. The hydrogen gas enters the bare-steel side of the test panel, passes through the steel, and exits out through the enamel when enough hydrogen-gas pressure has built up under the enamel. Upon exiting, a small piece of the enamel pops off the panel, creating a visible fishscale. A photo of the test apparatus is shown below.

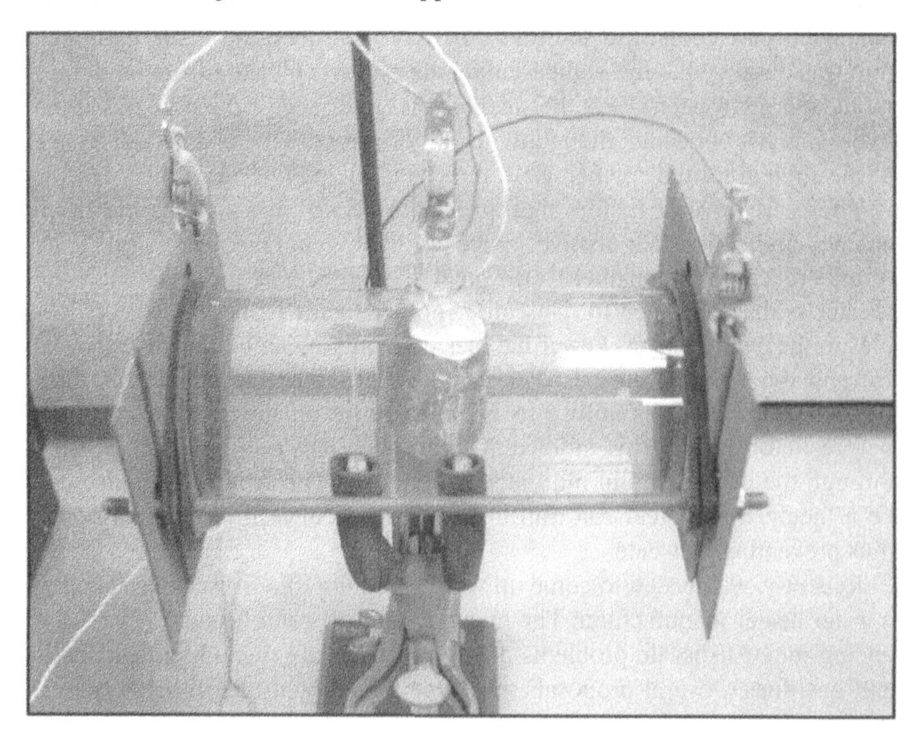

The number of fishscales that develop are recorded as a function of time. The "time to failure" is reached when the occurrence of each fishscale becomes so numerous that counting is impractical. In addition to testing the N-133, we also wanted to see the effect of changing the frit and clays in the mill addition.

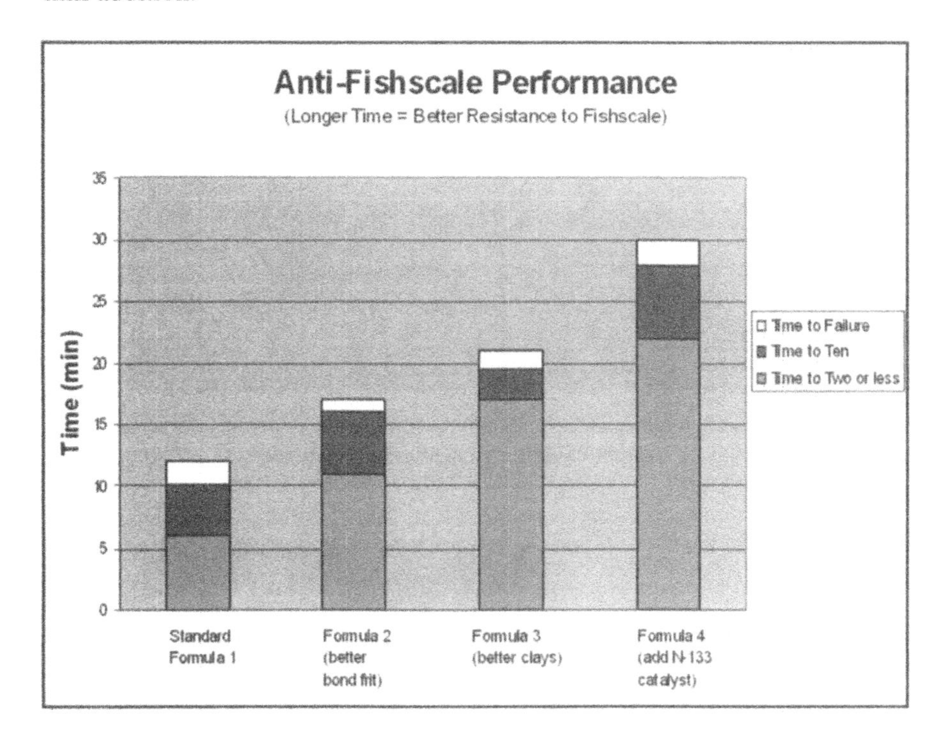

The chart above shows how the test can determine the relative fishscale resistance of the different experiments. The type of steel was the same in all the tests and, thus, was not a factor. Because of fewer bubbles in the enamel layer, formula 1, using clays that develop fewer bubbles is the least-resistant enamel. Substituting a better-bonding frit improves the resistance, but it is still hampered by the low-bubble clays (formula 2). Substituting other, larger-bubble, clays for these clays (formula 3) further improves the performance to about two times that of the original formula. Finally, the use of N-133 in the mill addition (formula 4) provides nearly a threefold increase.

In conclusion, by implementing all three steps—better bonding frit, better bubble structure, and addition of a nickel catalyst—fishscale can be greatly reduced, or even eliminated in a production environment.

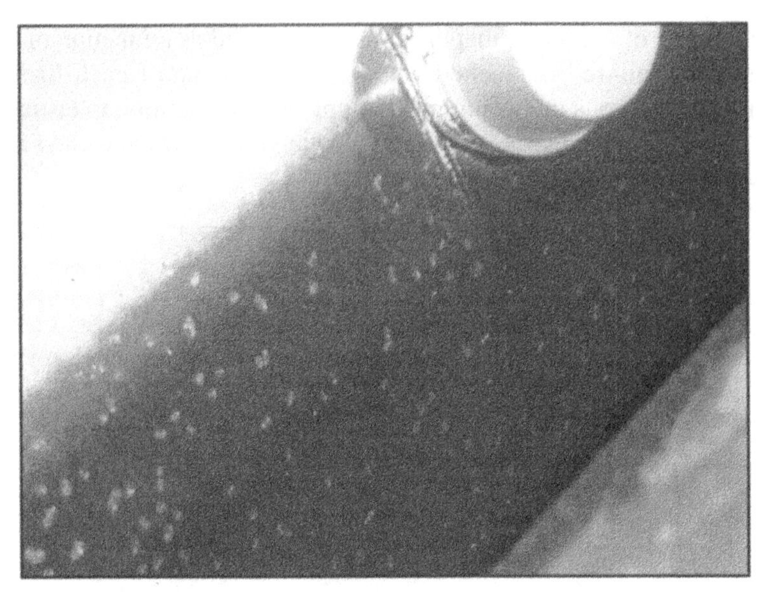

Factors Affecting Fishscale: enamel adherence, bubble structure of enamel layer, ability of enamel coating to chemically tie up atomic hydrogen, and type of steel.

Proposed Solutions: increase enamel adherence with a better bonding frit, change the enamel clays to increase bubble size, and use catalytic Nickel (N-133) to oxidize residual H_2 at steel/enamel interface to H_2O.

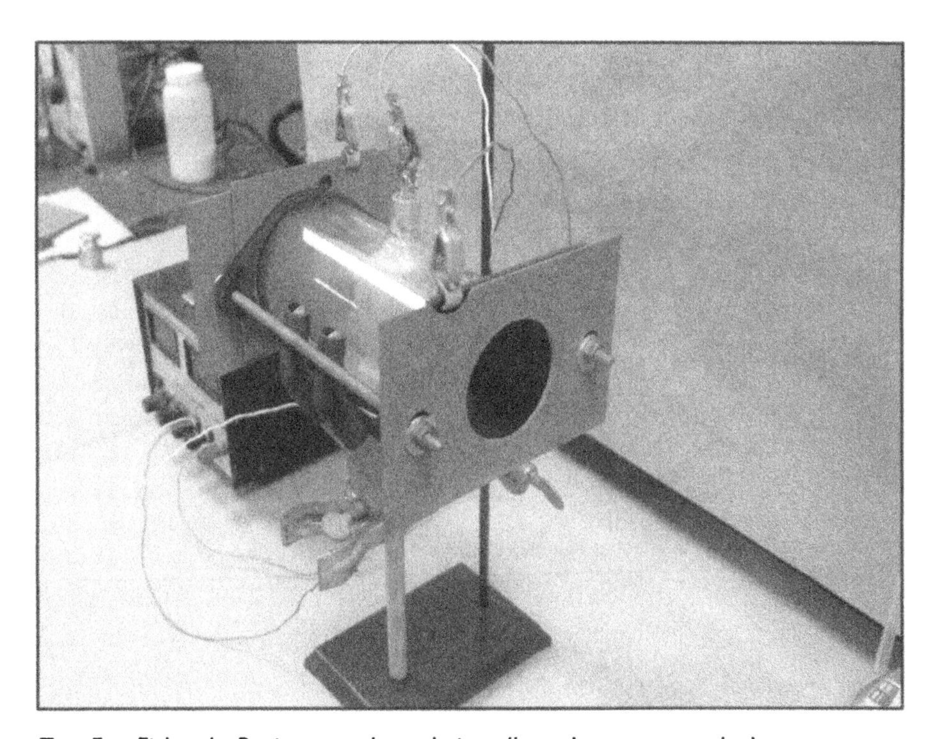

Test For Fishscale Resistance: electrolytic cell used to generate hydrogen gas, gas enters bare steel side of enamel coated plate, hydrogen gas builds up under enamel, causing fishscale, and number of fishscales that develop are recorded as a function of time.

Steel Qualification Process for Water Heater Tank Fabrication

Steve Sloan
American Water Heater Company

Abstract

Processes and procedures utilized to qualify steel suppliers and steel for use in fabrication of glass-lined tanks for water heaters are critical for efficient and cost-effective manufacturing in today's climate of escalating steel prices and limited supply and availability of materials. This presentation outlines basic principles developed over many years at American Water Heater Company.

American Water Heater produces gas and electric water heaters that are porcelain enameled and uses up to 250 tons/day of steel. During the past 6 years, up to 25 suppliers or types of steel have been evaluated. A three-step process has been developed and has been tested in more than 75 trials. It is a simple "common sense" approach. The basic principles are education, communication, and following procedures.

Purpose of the process:

- Establish clearly defined procedures to qualify steel and steel suppliers.

Objectives of the process:

- Establish efficient and timely steel qualification system.
- Reduce material costs.
- Reduce operating costs.
- Improve quality.
- Increase productivity.

Groundwork:

- Establish well-defined material specifications. (fig.1)
    1. Performance requirements, chemistry, mechanical properties, size, and packaging requirements.
    2. Industry standard specifications, such as ASME and ANSI.
- Establish well-defined trial and approval procedures. (fig.2)
    1. 200 piece trial.
    2. 2000 piece trial.

- Full coil, different heat (lot of steel).
- Define trial and testing protocol (visual analysis, surface condition, weld-ability, enamel defects, leak rates, enamel bond, hydrostatic, and pulsation testing). (fig.3)
- Documentation. (fig.4)

Basic principles:

- Education.

  1 First hurdle is to convince purchasing that steel is more than just price.

  2. Emphasize the sensitive nature of porcelain enamel and that minute changes in the steel substrate (chemistry, grain structure, etc.) may have devastating effects on the final fired enamel surface, as shown in the accompanying photos.

- Communication. (fig.5)
  1. Regular meetings.
  2. Published minutes.
  3. Assigned responsibilities.
  4. Action items.
  5. Schedules and timetables.
- Follow procedure.
- Team approach.

  1. Establish cross-functional teams typically from affected departments, including production, purchasing, and quality control, as well as suppliers of steel and porcelain enamel.

  2. Working together to be successful requires extensive coordination and planning of multiple operations, multiple departments, and multiple facilities.

  3. Remember the various goals, purchasing (lowest possible price), production (highest possible output), and quality (within specifications and satisfied customers).

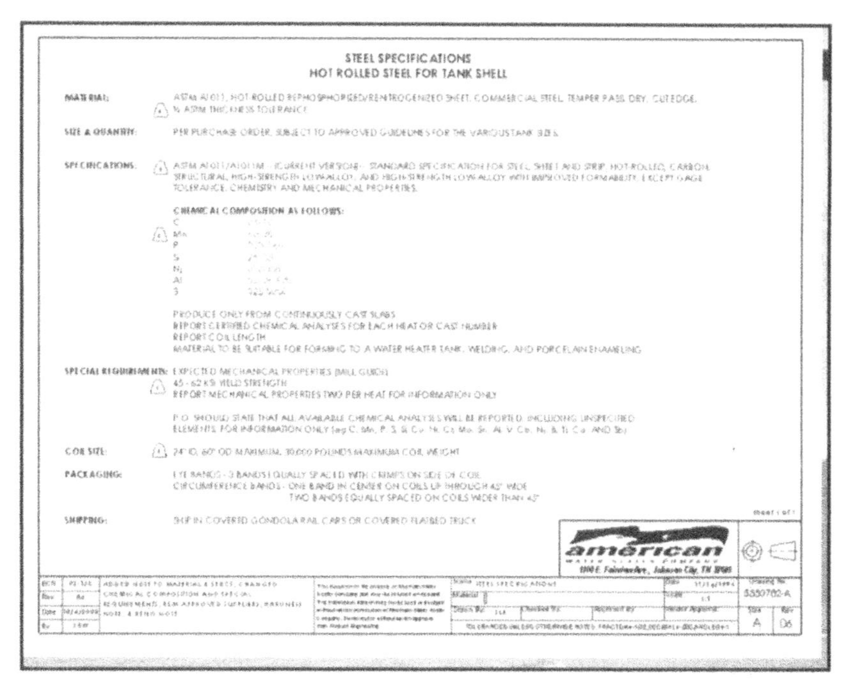

STEEL SPECIFICATIONS
HOT ROLLED STEEL FOR TANK SHELL

MATERIAL: ASTM A1011, HOT ROLLED REPHOSPHORIZED/RENITROGENIZED SHEET, COMMERCIAL STEEL, TEMPER PASS, DRY, CUT EDGE. ½ ASTM THICKNESS TOLERANCE

SIZE & QUANTITY: PER PURCHASE ORDER, SUBJECT TO APPROVED GUIDELINES FOR THE VARIOUS TANK SIZES.

SPECIFICATIONS: ASTM A1011/A1011M - (CURRENT VERSION) - STANDARD SPECIFICATION FOR STEEL, SHEET AND STRIP, HOT-ROLLED, CARBON, STRUCTURAL, HIGH-STRENGTH LOW-ALLOY, AND HIGH-STRENGTH LOW-ALLOY WITH IMPROVED FORMABILITY, EXCEPT GAGE TOLERANCE, CHEMISTRY AND MECHANICAL PROPERTIES.

CHEMICAL COMPOSITION AS FOLLOWS:
C
Mn
P
S
N
Al
B

PRODUCE ONLY FROM CONTINUOUSLY CAST SLABS
REPORT CERTIFIED CHEMICAL ANALYSES FOR EACH HEAT OR CAST NUMBER
REPORT COIL LENGTH
MATERIAL TO BE SUITABLE FOR FORMING TO A WATER HEATER TANK, WELDING, AND PORCELAIN ENAMELING

SPECIAL REQUIREMENTS: EXPECTED MECHANICAL PROPERTIES (MILL GUIDE)
45 - 62 KSI YIELD STRENGTH
REPORT MECHANICAL PROPERTIES TWO PER HEAT FOR INFORMATION ONLY

P.O. SHOULD STATE THAT ALL AVAILABLE CHEMICAL ANALYSES WILL BE REPORTED, INCLUDING UNSPECIFIED ELEMENTS, FOR INFORMATION ONLY

COIL SIZE: 24" ID, 60" OD MAXIMUM, 30,000 POUNDS MAXIMUM COIL WEIGHT

PACKAGING: EYE BANDS - 3 BANDS EQUALLY SPACED WITH CRIMPS ON SIDE OF COIL
CIRCUMFERENCE BANDS - ONE BAND IN CENTER ON COILS UP THROUGH 45" WIDE
TWO BANDS EQUALLY SPACED ON COILS WIDER THAN 45"

SHIPPING: SHIP IN COVERED GONDOLA RAIL CARS OR COVERED FLATBED TRUCK

sheet 1 of 1

american

5550702-A

Figure 1. Establish well-defined material specifications.

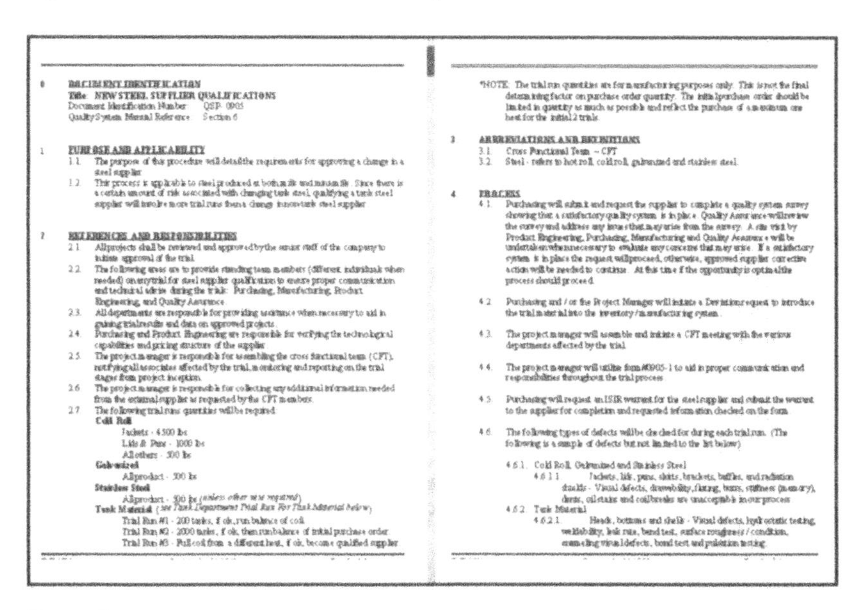

0 DOCUMENT IDENTIFICATION
Title: NEW STEEL SUPPLIER QUALIFICATIONS
Document Identification Number: QSP 0905
Quality System Manual Reference: Section 6

1 PURPOSE AND APPLICABILITY
1.1 The purpose of this procedure will detail the requirements for approving a change in a steel supplier.
1.2 This process is applicable to steel produced at both mills and minimills. Since there is a certain amount of risk associated with changing tank steel, qualifying a tank steel supplier will involve more trial runs than a change in non-tank steel supplier.

2 REFERENCES AND RESPONSIBILITIES
2.1 All projects shall be reviewed and approved by the senior staff of the company to initiate approval of the trial.
2.2 The following areas are to provide standing team members (different individuals when needed) on any trial for steel supplier qualification to ensure proper communication and technical advice during the trials: Purchasing, Manufacturing, Product Engineering, and Quality Assurance.
2.3 All departments are responsible for providing assistance when necessary to aid in gaining trial results and data on approved projects.
2.4 Purchasing and Product Engineering are responsible for verifying the technological capabilities and pricing structure of the supplier.
2.5 The project manager is responsible for assembling the cross functional team (CFT), notifying all associates affected by the trial, monitoring and reporting on the trial stages from project inception.
2.6 The project manager is responsible for collecting any additional information needed from the external supplier as requested by the CFT members.
2.7 The following trial runs quantities will be required:
Cold Roll
Jackets - 4500 lbs
Lids & Pans - 1000 lbs
All others - 500 lbs
Galvanized
All product - 500 lbs
Stainless Steel
All product - 500 lbs *(unless otherwise required)*
Tank Material *(see Tank Department Trial Run For Tank Material below)*
Trial Run #1 - 200 tanks, if ok, run balance of coil
Trial Run #2 - 2000 tanks, if ok, then run balance of initial purchase order
Trial Run #3 - Full coil from a different heat, if ok, become qualified supplier

*NOTE: The trial run quantities are for manufacturing purposes only. This is not the final determining factor on purchase order quantity. The initial purchase order should be limited in quantity as much as possible and reflect the purchase of a maximum one heat for the initial 2 trials.

3 ABBREVIATIONS AND DEFINITIONS
3.1 Cross Functional Team – CFT
3.2 Steel - refers to hot roll, cold roll, galvanized and stainless steel.

4 PROCESS
4.1 Purchasing will submit and request the supplier to complete a quality system survey showing that a satisfactory quality system is in place. Quality Assurance will review the survey and address any issues that may arise from the survey. A site visit by Product Engineering, Purchasing, Manufacturing and Quality Assurance will be undertaken when necessary to evaluate any concerns that may arise. If a satisfactory system is in place the request will proceed, otherwise, approved supplier corrective action will be needed to continue. At this time if the opportunity is optimal the process should proceed.
4.2 Purchasing and / or the Project Manager will initiate a Deviation request to introduce the trial material into the inventory / manufacturing system.
4.3 The project manager will assemble and initiate a CFT meeting with the various departments affected by the trial.
4.4 The project manager will utilize form #0905-1 to aid in proper communication and responsibilities throughout the trial process.
4.5 Purchasing will request an ISIR warrant for the steel supplier and submit the warrant to the supplier for completion and requested information checked on the form.
4.6 The following types of defects will be checked for during each trial run. (The following is a sample of defects but not limited to the list below)
4.6.1. Cold Roll, Galvanized and Stainless Steel
4.6.1.1 Jackets, lids, pans, skirts, brackets, baffles, and radiation shields - Visual defects, drawability, fluting, burrs, stiffness (memory), dents, oil stains and coil breaks are unacceptable in our process.
4.6.2. Tank Material
4.6.2.1 Heads, bottoms and shells - Visual defects, hydrostatic testing, weldability, leak rate, bend test, surface roughness / condition, enameling visual defects, bond test and pickling testing.

Figure 2.

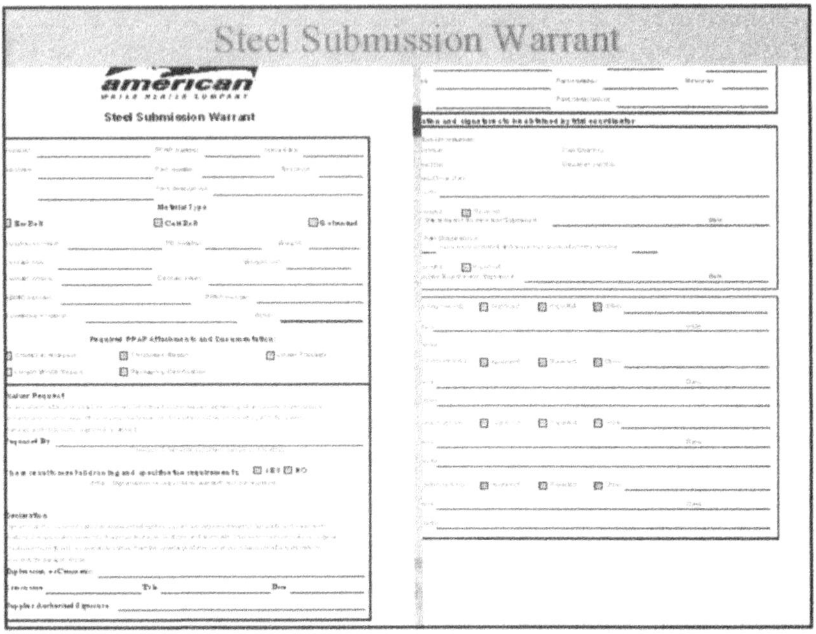

Steel Submission Warrant

american

Steel Submission Warrant

Material Type

Required PPAP Attachments and Documentation

Waiver Request

Requested By

Declaration

Supplier Authorized Signature

Figure 3.

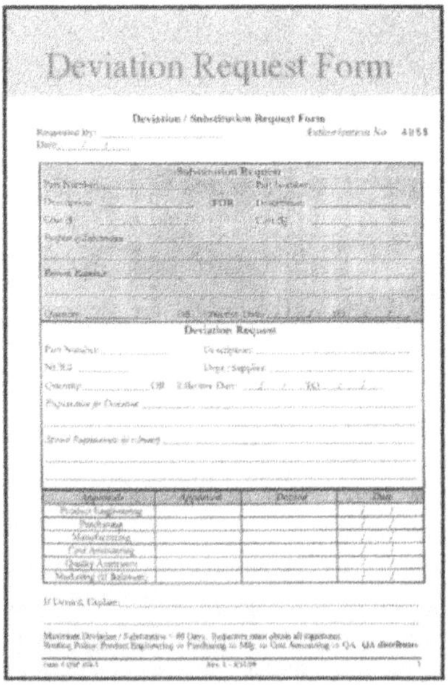

Deviation Request Form

Deviation / Substitution Request Form

Requested By: [illegible] No. 4855

Date:

Substitution Request

Part Number: Part Number:

Description: FOR Description:

Cost $: Cost $:

Reason of Substitution:

If Denied, Explain:

Quantity: OR Effective Date: TO

Deviation Request

Part Number: Description:

NCR# Dept / Supplier:

Quantity: OR Effective Date: TO

Explanation for Deviation

Special Requirements (if any)

Approvals	Approved	Denied	Date
Product Engineering			/ /
Purchasing			/ /
Manufacturing			/ /
Cost Accounting			/ /
Quality Assurance			/ /
Marketing (if Relevant)			/ /

If Denied, Explain:

Maximum Deviation / Substitution = 90 Days. Requestor must obtain all signatures.
Routing Policy: Product Engineering or Purchasing to Mfg. to Cost Accounting to Q.A. QA distributes

[illegible] Rev. 1 - [illegible] 1

Figure 4.

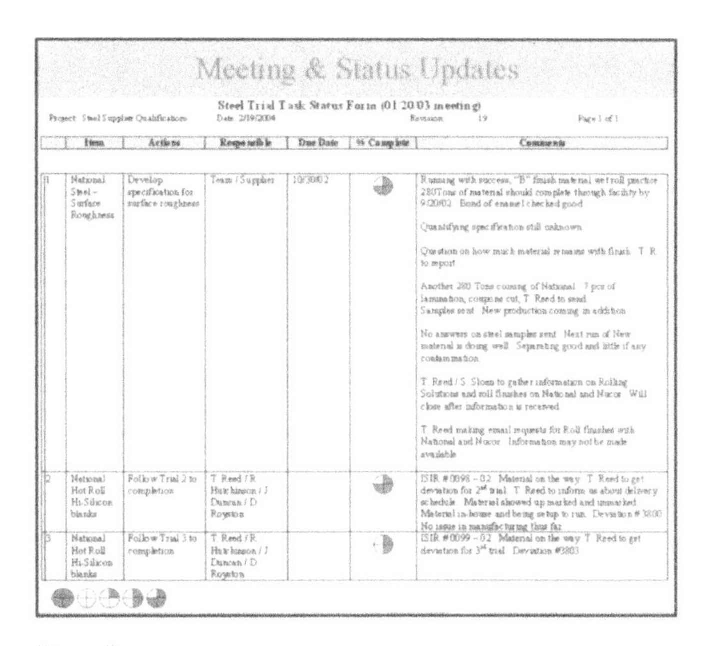

Meeting & Status Updates

Steel Trial Task Status Form (01 20 03 meeting)

Project: Steel Supplier Qualifications Date: 2/19/2004 Revision: 19 Page 1 of 1

	Item	Actions	Responsible	Due Date	% Complete	Comments
1	National Steel – Surface Roughness	Develop specification for surface roughness	Team / Supplier	10/30/02		Running with success, "B" finish material wet roll practice 280Tons of material should complete through facility by 9/20/02. Bond of enamel checked good. Quantifying specification still unknown. Question on how much material remains with finish. T. R to report. Another 280 Tons coming of National. 7 pcs of lamination, coupons cut, T. Reed to send. Samples sent. New production coming in addition. No answers on steel samples sent. Next run of New material is doing well. Separating good and little if any contamination. T. Reed / S. Sloan to gather information on Rolling Solutions and roll finishes on National and Nucor. Will close after information is received. T. Reed making email requests for Roll finishes with National and Nucor. Information may not be made available.
2	National Hot Roll Hi-Silicon blanks	Follow Trial 2 to completion	T. Reed / R. Hutchinson / J. Duncan / D. Royston			ISIR #0098 – 02. Material on the way. T. Reed to get deviation for 2nd trial. T. Reed to inform us about delivery schedule. Material showed up marked and unmarked. Material in-house and being setup to run. Deviation # 3800. No issue in manufacturing thus far.
3	National Hot Roll Hi-Silicon blanks	Follow Trial 3 to completion	T. Reed / R. Hutchinson / J. Duncan / D. Royston			ISIR #0099 – 02. Material on the way. T. Reed to get deviation for 3rd trial. Deviation #3803.

Figure 5.

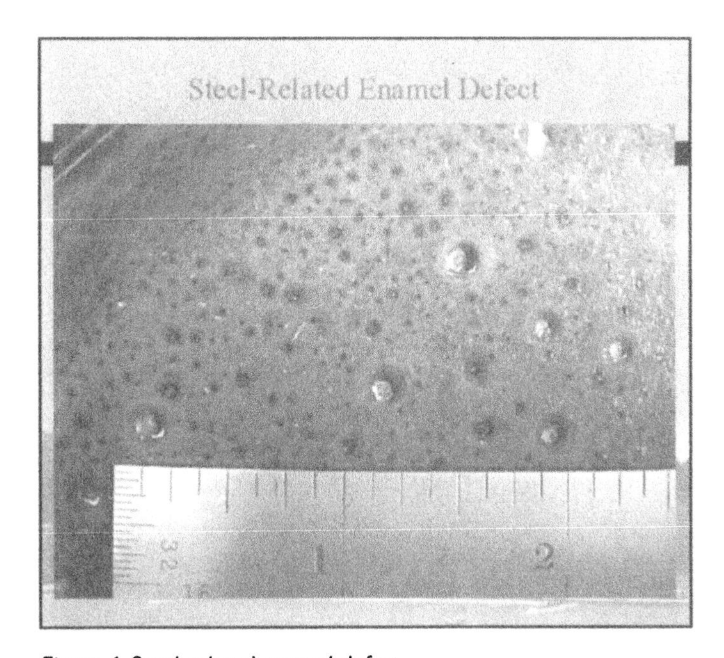

Figure 6. Steel-related enamel defect.

Quantifying Bubble Structure in Water Heater Enamels

Len Meusel
Pemco Corporation

Phillip Stevens
Rheem Manufacturing

Introduction

The art of enameling water heaters has been retained in the experience of a handful of skilled professionals for years. For water heater applications, a good bubble structure is an integral part of the enameling system. The evaluation of the bubble structure for these enamels always has been subjective. Usually defined as good or bad, dense or sparse, underfired or overfired, it becomes more qualitative than factual, quantitative, and objective.

The purpose of this paper is to try to determine if it is possible to demonstrate an objective way to quantify the evaluation of water heater enamels in respect to bubble structure.

In this paper, bubble structure will be defined as a combination of bubble population and density. Bubble population refers to the sizes of bubbles in the enamel, while density is the ratio of the glass to space occupied by the bubbles.

Background

In an ideal world, there would be no bubbles in enamel glass. This would provide the perfect water resistance. But, because we need to be able to apply the enamel, a formulation including clays and electrolytes is necessary. The bubble structure created in the enamel is the product of the slurry (clay and electrolytes) and the outgas from the substrate. The clay in the formula creates the basic bubble size and density, and the electrolytes, substrate, and glass modify it. The viscosity of the glass also determines the size of the bubbles. The glass needs to be viscous enough to retain the small bubbles and fluid enough to allow the larger bubbles to coalesce and escape. Essentially, the glass and the slurry need to be formulated for proper rheology and for the application process and firing profile of the customer.

The bubble structure of water heater enamel serves many purposes. It provides a place to store hydrogen. This is important, because most water heaters are fabricated out of excessively gaseous hot-rolled steel. Therefore, the capability of the bubble structure to absorb hydrogen produces a smooth surface and inhibits fishscale. The bubble structure makes the enamel coating slightly flexible. This flexibility is necessary, because a water heater flexes as it heats and cools during normal operation. Once the average bubble structure is established in a production facility, it can be used as a tool to control quality and processes. Variations in the bubble structure could show irregularities from proper production in firing temperature and application weights, as well as misweighed mill additions or improper setup.

Therefore, what bubble structure is desired by the water heater manufacturer to obtain optimum results? What sizes are small, medium, and large bubbles? How big is too big? What is good density? What other enamel properties are related to bubble structure? These are the questions we are going to try to quantify through measurement and photomicrographs.

Analysis of Bubble Structure

Proper analysis of water heater enamel bubble structure must begin with proper application and firing of the enamel. The average fired thickness should be 7–9 mils or 158–230 μm. Furnace temperature should be at regular production settings, and tanks should be loaded at normal capacity.

Examination of the bubble structure is usually done at 100 magnifications under a high-power microscope. In order to quantify the size of the bubbles, several pieces were examined to determine various sizes. The following measurements represent averages of the different bubble sizes when photographed at 100 magnifications: very small, <15 μm; small, 16–25 μm; medium, 26–35 μm; large, 36–65 μm; extra-large, 66–99 μm; and oversize, >91 μm.

Density

Density, as stated earlier, is the ratio of the glass to the space occupied by the bubbles. It is the number of bubbles in a given area. The population and size of the bubbles in the fired enamel coating determine good bubble structure.

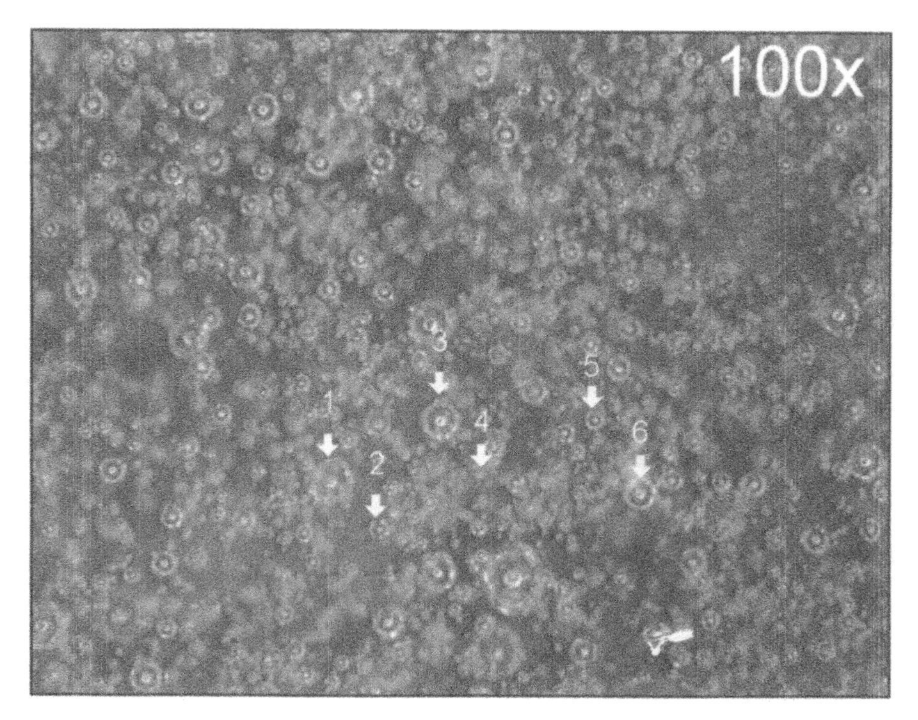

Figure 1. Good bubble structure ((1) 47 μm, 1.85 mils, extra-large; (2) 17 μm, 0.66 mils, small; (3) 31 μm, 1.22 mils, medium; (4) 8 μm, 0.31 mils, very small; (5) 16 μm, 0.62 mils, small; and (6) 31 μm, 1.22 mils, medium).

Good Bubble Structure

Ideal bubble structure consists of small- to medium-sized bubbles with good density. A good bubble structure also could consist of a high density of small bubbles or a slightly less dense population of medium-sized bubbles.

The following scenarios depict variations of bubble population, evenly distributed for good density, which would qualify as excellent bubble structure. Refer to Figures 1 and 2 for examples. There also are guidelines that must be taken into consideration for the hot-roll-steel substrate. Where this steel is used, a few oversized bubbles are going to be normal when mixed with a good density of small to medium bubbles.

(1) Small bubbles comprising 30%–50% of the enamel;

(2) Small to medium bubbles comprising 30%–50% of the enamel;

(3) Medium bubbles comprising 30%–40% of the enamel;

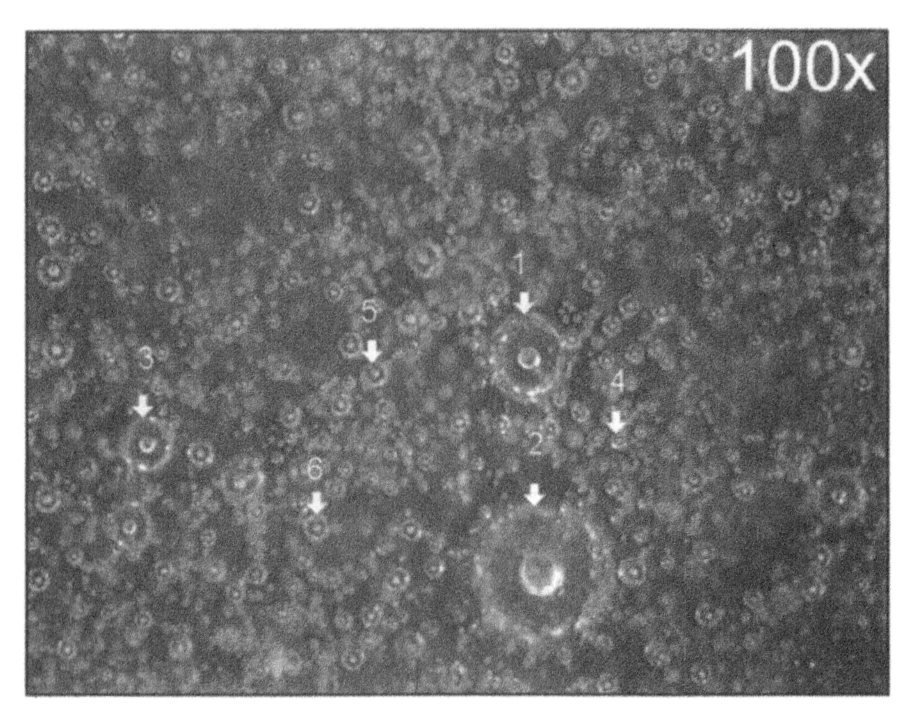

Figure 2. Good bubble structure ((1) 75 μm, 2.95 mils, extra-large; (2) 139 μm, 5.47 mils, oversize; (3) 45 μm, 1.76 mils, large; (4) 15 μm, 0.58 mils, very small; (5) 19 μm, 0.75 mils, small; and (6) 21 μm, 0.84 mils, small).

(4) Small, medium, and large bubbles comprising 30%–40% of the enamel; and

(5) Medium to large bubbles comprising 25%–35% of the enamel.

Acceptable bubble structure could be characterized by parameters.

(1) Small bubbles comprising 25%–55% of the enamel;

(2) Small to medium bubbles comprising 25%–55% of the enamel;

(3) Medium bubbles comprising 25%–45% of the enamel; and

(4) Small, medium, and large bubbles comprising 20%–40% of the enamel.

Poor bubble structure is any enamel with bubble population falling outside the above parameters. Figure 3 illustrates a good example of poor bubble structure.

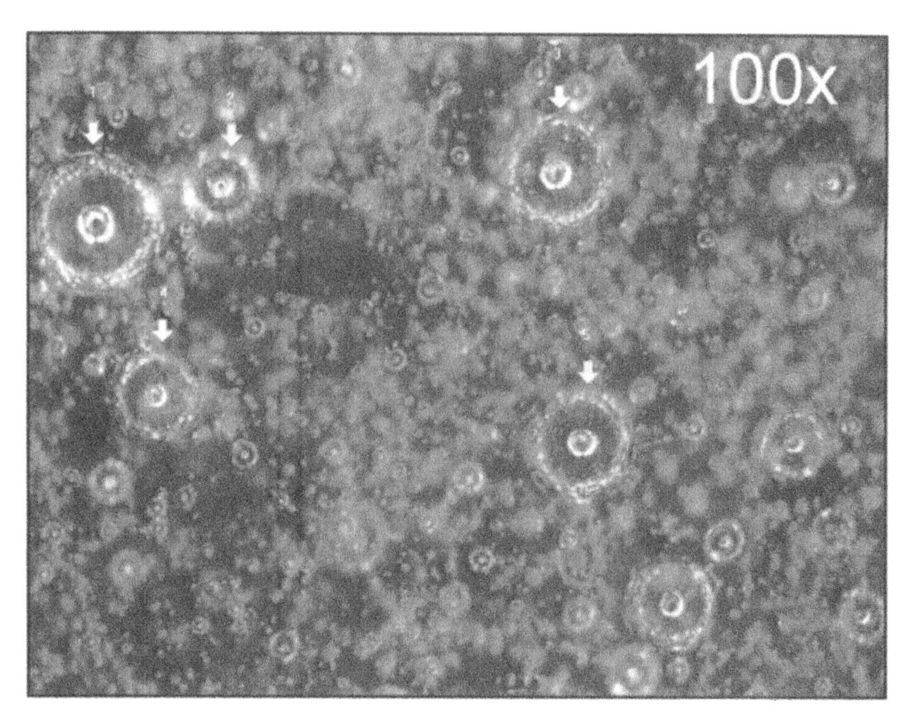

Figure 3. Poor bubble structure ((1) 123 μmm, 4.8 mils; (2) 56 μmm, 2.2 mils; (3) 94 μmm, 3.7 mils; (4) 84 μmm, 3.3 mils; and (5) 99 μmm, 3.9 mils).

Enamel Properties Related to Bubble Structure

It is important to understand the importance of the relationship between bubble structure and the enameling process. The following sets of photomicrographs depict conditions that indicate irregularities in the enameling process because of the variation in the bubble population and density. Figures 4 and 5 show an underfired condition. The underfired bubble structure is much denser and has a greater number of small bubbles. This is normally due to the part not getting enough time or temperature while in the furnace. This also could be caused by heavy application weights or too much silica added to the mill. A change in substrate thickness also would draw heat as well as an increase in the load in the furnace.

The overfired condition illustrated in Figures 6 and 7 show the opposite. The enamel is less dense with many voids, fewer small bubbles, and more medium to large bubbles. If the furnace settings are correct, this condition would indicate a misweigh, usually meaning the silica was probably left out

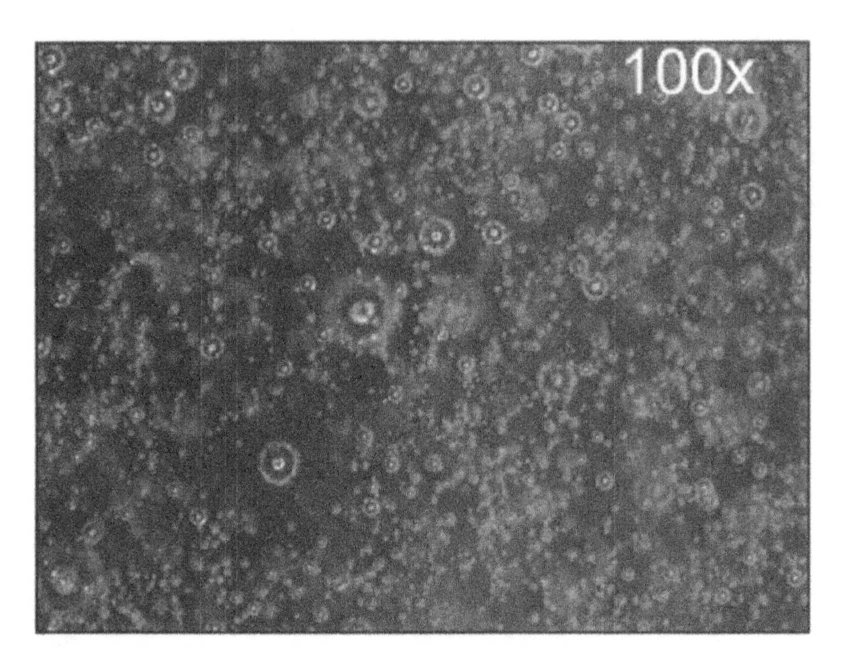

Figure 4. Good bubble structure.

of the mill. The surface also would be glossier.

Improper addition of set-up salts usually results in the excessively large bubbles, as illustrated in Figures 8 and 9. The structure exhibits good density, but the overall size of the bubbles is mostly large to extra-large. Other possible causes could be localized organics on the substrate prior to enameling, or a strong salt was left out of the mill.

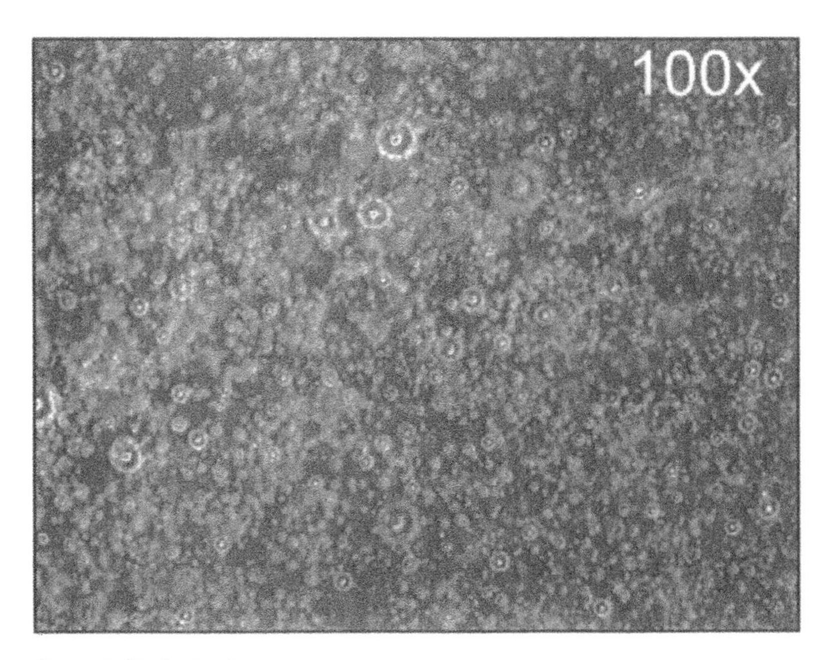

Figure 5. Underfired.

Conclusion

The goal of this paper was to demonstrate a quantitative method of distinguishing good and poor bubble structure for water heater application. What we have accomplished is a better means by which the water heater novice can understand bubble structure based on bubble population and density. What we determined was that it is still subjective, because there are so many scenarios of good bubble structure. It will be the responsibility of customers with the help of their enamel supplier to determine what the best bubble structure for their application should be and maintain that bubble structure through proper enameling practices.

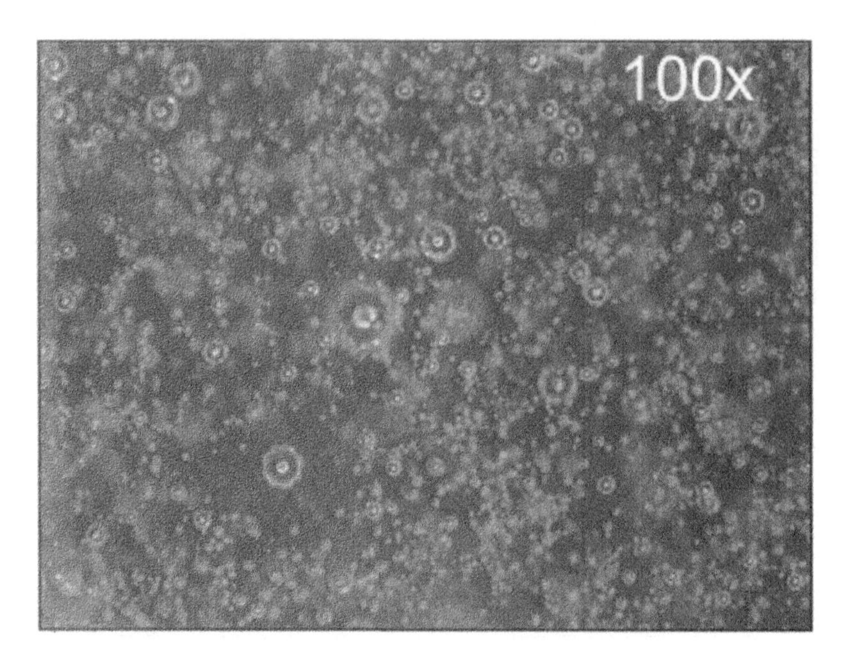

Figure 6. Good bubble structure.

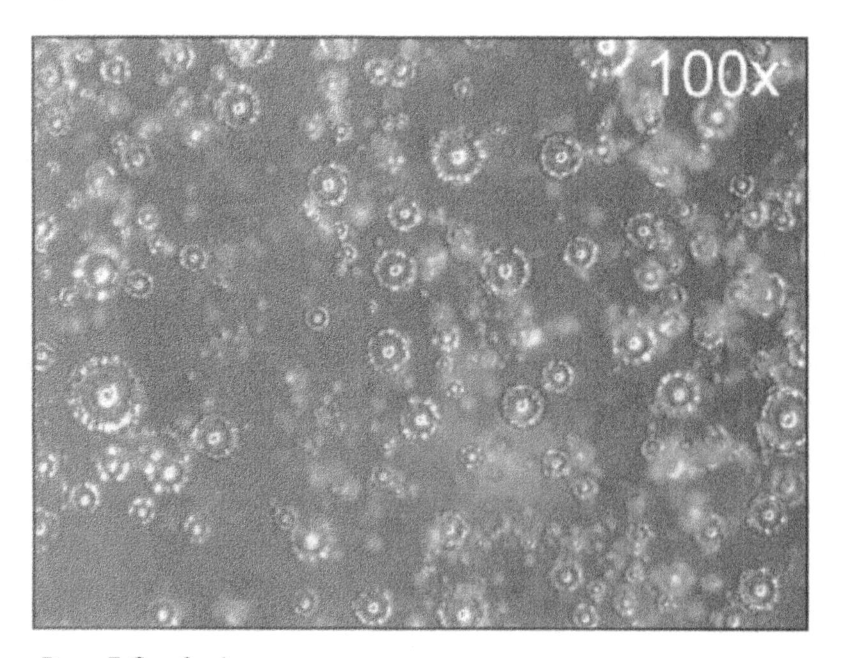

Figure 7. Overfired.

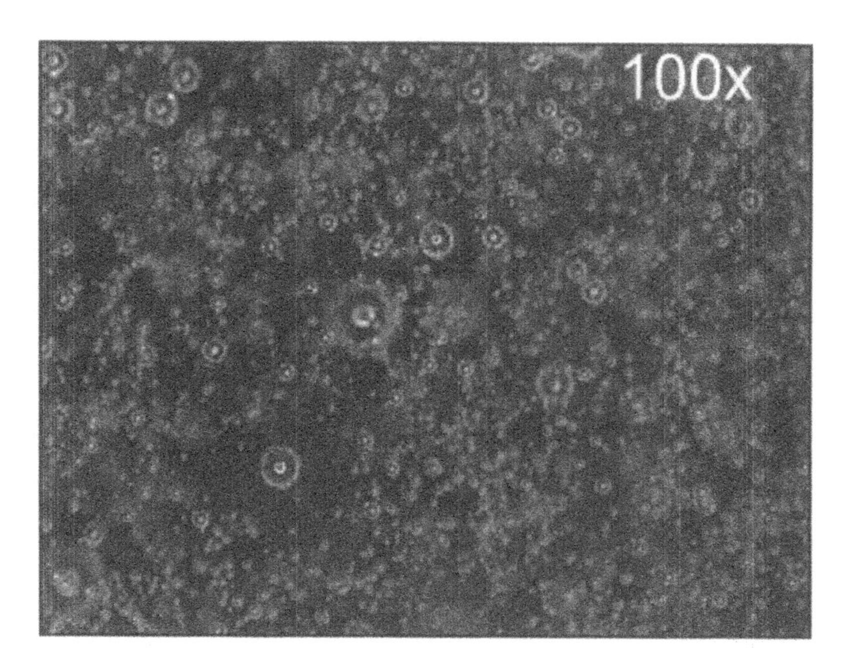

Figure 8. Good bubble structure.

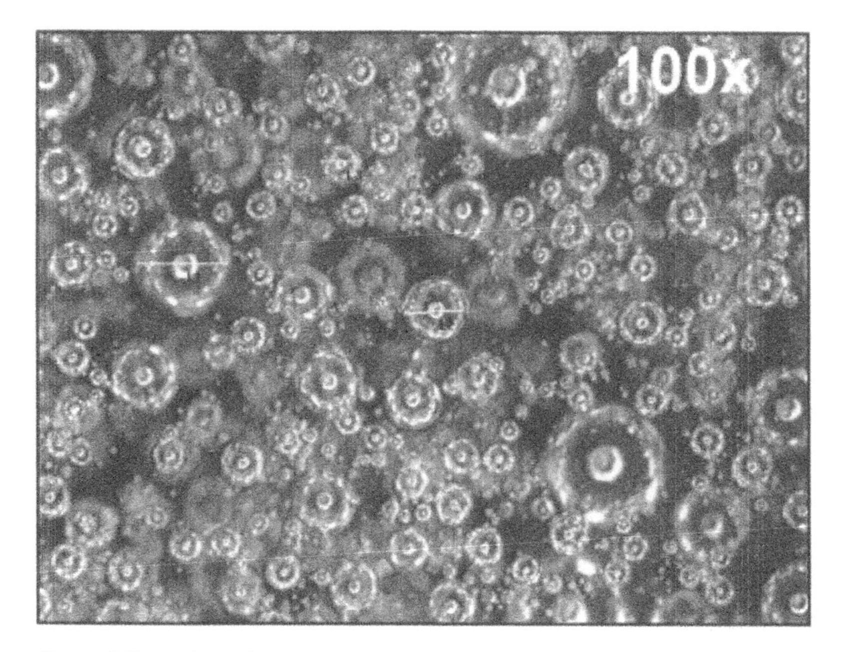

Figure 9. Excessive salts.

Boiler/Water Heater Inside Coating with Wet Enamel

Hans-Jürgen Thiele
E.I.C. Group GmbH, Member of DEV, Dietzenbach, Germany

Introduction

This paper will show and explain the wet-enamel application technique with flooding and low-pressure-spraying systems that is currently used. The main points of interest are the general concept of systems and the process technique used for bigger manufacturer series.

It could be asked if the quite popular wet-coating process is still an economical solution in comparison with the powder-coating process.

Usually, manufacturers of electric hot-water heaters use the wet- and powder-coating processes for small production rates. However, when changing from wet to powder, it is often necessary to adapt the boiler construction and shape to the powder-coating process.

Can the wet-coating process still be considered an actual application process compared with the powder-coating systems installed in the recent years?

Surely, the answer is yes, because of

- the generally increasing production rates for water heaters with internal pipes;
- change of production from stainless steel to normal steel;
- modifications in boiler construction;
- increase of boilers above 100 L volume;
- new wet application techniques with increased automation concepts; and
- the necessity of producing a higher quality among competitors.

This has led to multiple investments in new wet-coating lines since the end of the 1990s. These investments were planned according to actual economical and ecological aspects.

In the past five years, boiler-coating lines with wet enamel have been installed in Europe, North America, and North Africa in accordance with the several criteria.

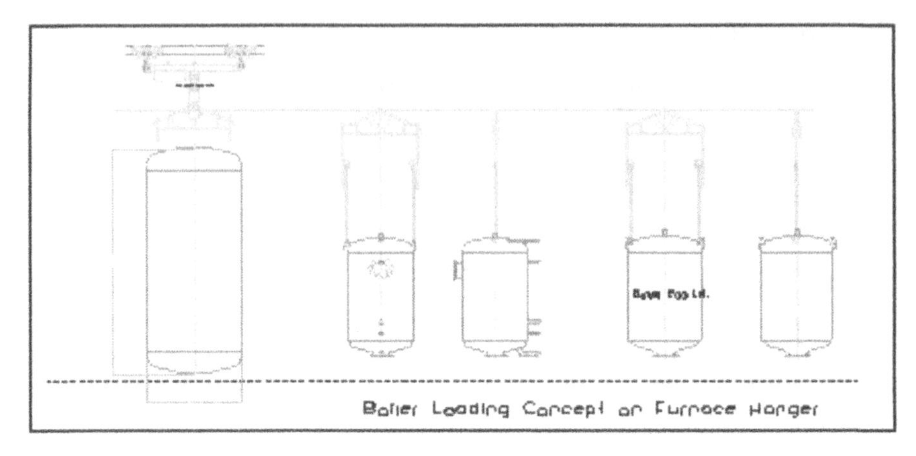

Figure 1. Boiler-loading concept on furnace hanger.

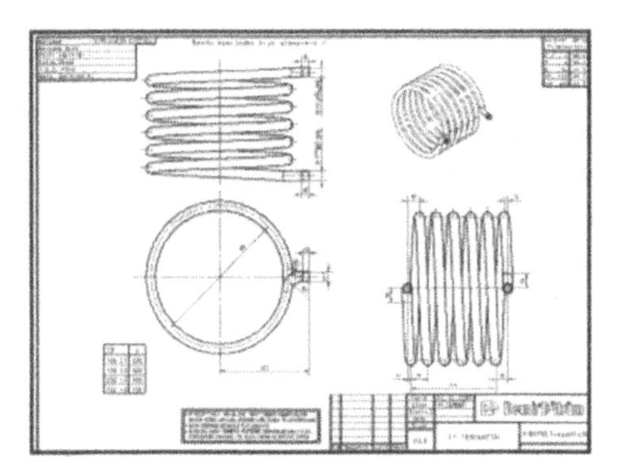

Figure 2. Schematic diagram of coil tobe enamelled.

The Economical Aspect

The demand for the highest flexibility in a technical solution, which is requested for electrically heated water heaters as well as for those with pipes inside (heat exchanger), was forced by the following aspects: (i) the rate of using water heaters with internal pipes increased to ~70%; (ii) the water heater size now is between 80 and 500 L, with an average between 200 and 300 L; (iii) the minimum production capacity raced above 800 boilers/(7.5 h shift) (some installations have an average of more than 1500 boilers/(7.5 h shift)); and (iv) the automation of coating process according to the wide range of boiler shapes as well as the diameter and the length generally required and installed. Also, the variation of pipe shapes with the various outlet pipe connections does have an extreme effect on the design of the system solution.

The degree of automation for the entire production process is based on the following: (i) formation possibilities direct from coil; (ii) welding immediately upon formation; (iii) the following pretreatment 1× or 2× degreasing and 2× rinsing with spraying nozzles and water drier; (iv) enamel-coating process by spraying or flow coating with drier; (v) a manual cleaning of pipe

outlet connection; (vi) the furnace design; (vii) a floor or overhead transport to and from individual system process stations are automated; and (viii) enamel slip circulation is now in a closed circuit.

The Ecological Aspect

The ready-to-use enamel delivered by various enamel manufacturers comes in big bags, which avoids the use of mills and considerably improves the environmental aspect of the wet application. Similar to powder, wet enamel is used in closed circuit.

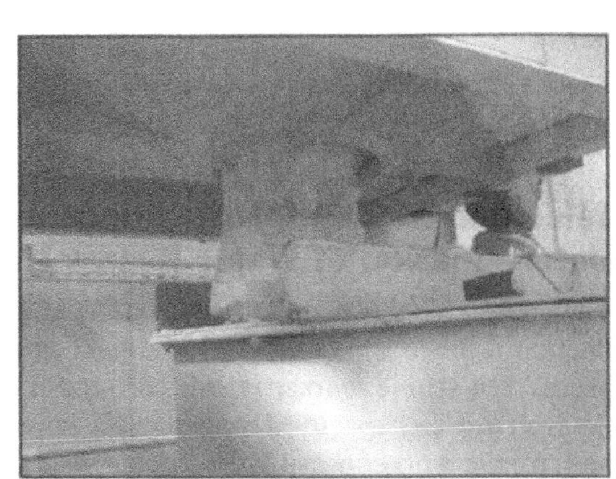

Figure 3. Big-bag system for ready-to-use process.

Figure 4. Photograph of hanger-loading concept.

Selection Criteria for the Application and Process Determination of Water Heater Manufacturing

The entire global concept of the actual coating lines designed for high production is interesting. Most of the tasks are running fully automatically, except that (i) the worker is still cleaning the pipe outlet connections, (ii) the operator sets specific parameters for coating of the inside of the boiler, and (iii) the worker possibly loads/unloads the boilers.

The remaining process steps are fully automated:

- Water heaters are mostly transported automatically from the welding to the application plant with the help of an overhead conveyor. The loading can be either at a zero point position up or down. The opening for entering with lances or spraying guns can be in an up or a down position. Main advantages are that the water heaters in new plants are loaded from the beginning directly to the furnace conveyor. This means that the water heaters are loaded at the welding and unloaded stations after firing. Also, if a transfer from one line to another is needed, it can be done automatically.
- Contrary to former pretreatment techniques with a dipping process, currently all actual investments for pretreatments are designed for

spraying with following steps: (i) 1× or 2× degreasing at 60°C–70°C during ~90–120 s; and (ii) 2× rinsing with cold water during ~50 s. Also, an eventually needed following pickling with an attached rinsing is effected using a spraying process and 1× passivation at 40°C–50°C during 110–120 s, all by spraying with nozzles. Also, the blasting technique is still in use if the internal pipe design does allow for blasting.

- After the wet pretreatment, the water heaters are dried using a drier with hot air at 120°C–140°C during 8–10 min.
- The wet application by automatically driven flow coating is done at a water heater position between 30° and 110°. The coating happens during the rotation of the water heater.
- The slip transfer into the water heater requires pumps with high fluid volume transfer in the shortest time.
- The production time for one water heater transport, swivelling, flooding, draining, and distribution is dependent on the process

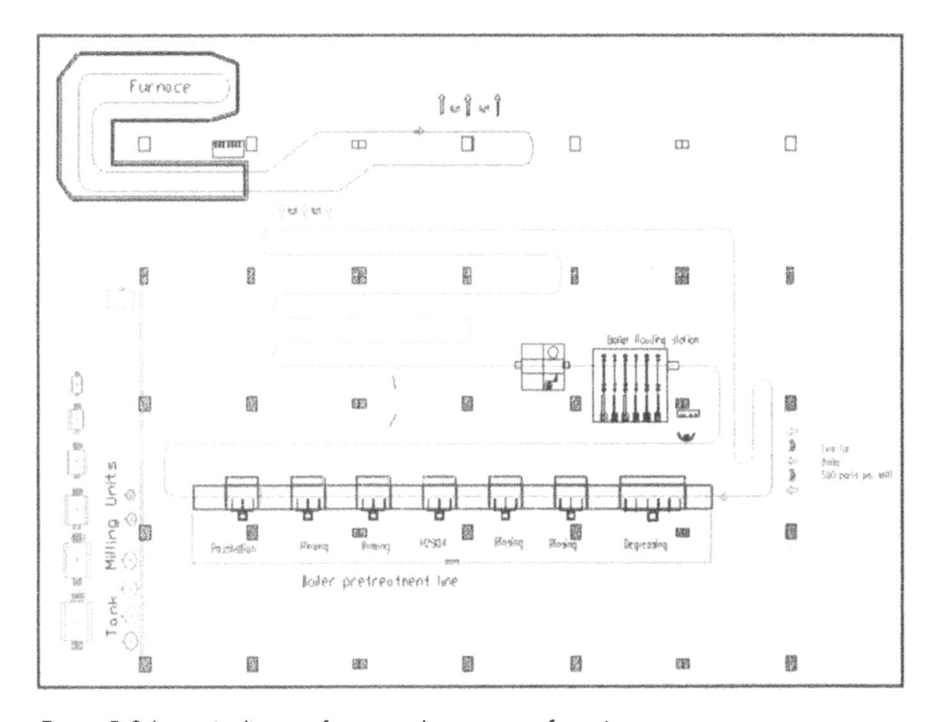

Figure 5. Schematic diagram for water heater manufacturing.

time required, as determined by (i) the water heater size; (ii) diameter of heating pipes, position, construction, and distances of pipes; (iii) the slip material adjustment; and (iv) required number of coatings for thickness and quality. The spreading out and distribution of the enamel onto the water heater inside surface is affected—depending on the system capacity—after the flooding process or within two separate one or two turning stations following the flooding stations. In order to do so, the water heater is moved and rotated—in accordance with the internal pipes/heat exchangers position—to the diagonal positions as programmed.

- The outside coating (if required) can be applied directly after the inside coating.
- Air nozzles with heated air (120°C–140°C) will dry the inner area of the water heater during ~8–10 min.
- Before firing, the applied inner surface is controlled manually.
- The outside flange and pipe connections are cleaned manually in most cases.
- Depending on the capacity and customer requirements, the transfer to the furnace conveyor is affected automatically or manually.
- The firing process is accomplished.
- The water heaters are automatically or manually unloaded from the furnace conveyor.

The main features for the slip transfer of the actual automatic wet flow coating systems of water heaters are as follows:

- There is constant filtration and circulation in a closed circuit of the slip.
- The slip is automatically proportioned according to the water heater size and its pipe configuration in order to coat the inside and the pipes properly.
- The working tanks are equipped with flow level control sensors that will call for fresh material, which is transferred to the tank with diaphragm pumps.
- In addition, pipe surfaces that are very difficult to reach also are evenly applied with the enamel coating material.
- The slip can be set up in accordance with the individual requirements.

- A constant slip consistency offers a regular thickness and surface quality.

All the above-mentioned criteria have been taken into consideration for the plant design with the individual customer during the planning of the water heater coating system.

The Water Heater Inside Coating with Low-Pressure Spraying Technique

In general, E.I.C. has developed low-pressure spraying systems that are exclusively used by the manufacturing industry for products to be enamelled.

Our focus is based and concentrated on (i) the improvement of the surface quality, (ii) a perfectly automated production run from the loading of the conveyor to the unloading for the assembly, (iii) highly efficient transfer of the enamel material, and (iv) a reduction of wear of parts that are in contact with enamel, e.g., nozzles, needles, pumps, and valves.

Also, for this application process, the water heaters are transported with an overhead conveyor through a spraying pretreatment and dryer.

In contradiction to the flooding technique—in which the water heater is pretreated, coated, and fired at the furnace hanger—the water heater is now taken from the pretreatment line and transferred to the coating line and then transferred again to the firing line.

Independent of this transfer requirement, there is an advantage of a higher manufacturing output. More than 1500 boilers/(7.5 h shift) can be enamelled.

The low-pressure spraying technique—with material pressure of ~0.2–0.3 bar (3.0–5.0 psi) and an atomization pressure of ~1.0–2.0 bar (15–30 psi) at the spray gun head—offers the advantage of a high material transfer efficiency and a low overspray. The inner surface thickness of the water heater is between 150 and 220 μm.

This application technique is based on enamel slip circulation via the spray gun head and back into the working tank, which allows the mentioned low material pressure of ~0.2–0.3 bar inside the spray gun head. This circulation process (the slip is always in motion and enamel cannot settle) avoids clogging inside the hoses and achieves constant manufacturer parameters.

Additionally, the cleaning system, developed by E.I.C., returns the remaining slip, still resting in the material hoses, back to the working tank after production is finished or for color change.

Figure 6. Flow coating system.

The flow rate at the spray guns can be individually adjusted for each spray gun in accordance with the customer requirement.

To allow the inside spraying, the water heater is separated in two parts: (i) the tank shell with welded top; and (ii) the bottom with and without flue pipe.

Depending on the different parts shapes and in order to achieve an ideal automatic application process, the parts will be coated in two different systems.

- System technique for tank shells with a top: (i) tank shells are loaded with the opening up-side-down on an overhead conveyor; (ii) each tank will be detected before entering the spray booth; (iii) the automatic spraying guns do not spray if hangers are not loaded; (iv) the tank shells are positioned automatically to the spray location inside the spray booth; (v) when the parts are in the spraying position, the hangers will start spinning, and the guns are moved up into the tank shell; (vi) to spray the upper area inside the tank, the guns are automatically moved from horizontal to vertical spraying position; (vii) the speed of the stroke and the spinning frequency of

Figure 7a. Coating outsides of tanks.

Figure 7b. Coating outsides of tanks (con't).

hangers are adjustable; and (viii) after the tank shells are sprayed, the tank fixation assembly opens and the conveyor moves one step forward. Four tank shells can be sprayed at the same time.

- The flue bottom with and without flue pipe: (i) flue bottoms are carried on a floor conveyor double-track system; (ii) parts are detected before entering the spray booth; (iii) fixed and movable spray guns will coat the flue bottom outside surface; (iv) in this installation, three parts are enamelled on an indexing basis; (v) following the coating, the weld area is automatically cleaned of enamel with air nozzles; and (vi) servo motors are automatically positioning the air nozzles depending on the bottom ring diameter.

The E.I.C. automatic gun operates always with a continuous enamel circulation through the gun head. This system avoids blockages that are normally frequent in spraying enamel out of pressure feed tanks or without a circulation system. The enamel circulation works similar to a hydraulic principle. The enamel output depends on the pressure difference between enamel in and the output of the gun. The material flow will be activated by opening the needle position and the fluid tip passage at the gun. The material output can be regulated by increasing or decreasing the material pressure/speed of the fluid material inside the hoses.

Figure 7c. Coating outsides of tanks (con't).

The Slip Supply System

The slip supply system from the storage tank placed inside the mill room to the spraying guns transfers the material automatically to the material tank off at the spraying stations. All storage tanks are equipped with fluid level controls. When the minimum level is reached, a signal is given to the main control panel to refill a tank or to transfer material to another tank. Each tank is equipped with an electrically driven agitator to avoid material settlement of the slip inside the tank. Mixing times of the slip material is programmable in order to keep the slip ready to use. Each tank has two outlets: one deeper than the bottom level to empty and clean the tank; and the other one for the enamel supply to the working tank.

Enamel Supply System from the Working Tank to the Spraying Guns

The enamel supply from the working tank to the spray system also is run with diaphragm pumps with a pulsation chamber in order to avoid pulsation inside the hoses. The pumps will work continuously and without interruption. They forward the enamel to the guns through the gun head and back to the tank.

Each pump on/off mode is controlled from the control panel. The material amount can be adjusted by the material pressure (which is not really precise) or (more precisely and better) by the stroke frequency of the pump.

Material Supply Cleaning System

Cleaning means a time-controlled pulsing on two valves. Our water/air cleaning system has been designed for reusing effectively the slip inside pumps and fluid hoses. To avoid any risks of production stoppage, we recommend the use of this cleaning process between two working shifts as well as daily production, or during production stops of longer than 20 min. Duration of the cleaning process is no longer than 4–5 min.

Customer Benefits

There are several customer benefits with the E.I.C. material-saving intelligent spraying.

- Photocell/computer-controlled gun-triggering system: (i) the gun will spray only if there are parts loaded to the hanger (advantage: lower production cost); (ii) the guns will be triggered individually according to the height and length of the parts (advantage: optimal and quick adaptation to the part dimension); (iii) because of lower overspray and controlled triggering, there will be a smaller quantity of reclaimed enamel inside the cabin (advantage: enamel material saving and less cabin cleaning); and because of less reclaimed enamel, mixing ratio of fresh enamel with reclaimed enamel will be higher (advantage: better quality of sprayed enamel, better surface quality, and reduction of waste disposal costs).
- Patented low-pressure spraying atomizer: (i) very low atomization pressure needed (advantage: less air atomizing and less air energy as well as spray pattern and coated surface are more even and smoother, i.e., less orange peel effect); (ii) during atomization, very low bounceback from the part (advantage: the exhaust air speed inside the spray booth can be reduced by nearly 50%, which saves energy); and (iii) very low material pressure and fluid speed at the nozzle (advantage: the lifetime of the nozzle is much longer (minimum of 6 months, whereas usual nozzles will wear-out after 6 weeks) as well as production loss and maintenance times are less).

- The dual material supply and circulating system is the most important part of this intelligent spraying technique: (i) automatically controlled fluid slip transfer from the mill room to the individual working tanks (advantage: no staff needed for slip transfer via forklift); (ii) working tanks with integrated fluid level control (advantage: smaller working tanks that can be cleaned quicker and with less water consumption as well as slip saving because of less surface contact); and (iii) slip circulation (advantage: no production interruption or losses from settlement in the hoses and spray guns).

Summary

When choosing and determining a new system technique, the most important criteria are the following:

- A quick adaptation of the production system to the constructive changes of product to be coated without having high and cost-intensive new investments;
- High grade of automation;
- Flexibility in enamel adjustment; and
- Reduction of ecological damage.

These criteria are the parts and deciding factors for the water heater systems, installed in the past years, and will also be for new enamelling systems.

RealEase™ Nonstick Porcelain Enamel

Charles Baldwin, Alain Aronica, Brad Devine, and Graham Rose
Ferro Corporation

RealEase™ is an innovative, ceramic-based nonstick coating developed by Ferro. It has excellent scratch, abrasion, and heat resistance as well as superb cleanability.

Introduction

Because it is a glass-based coating, porcelain enamel is much more scratch, abrasion, and heat resistant than organic paints. However, burned-on food residue forms hydrogen bonds and strongly adheres to enamel.[1] The cleanability of enamel can be improved by maximizing the acid resistance, catalyzing the transformation of burn-on residue into ash, pyrolyzing the residue into ash,[2] or applying an easy-to-clean top-coat to the enamel.[3,4] However, none of these have the cleanability of organic nonstick coatings.

The two major families of organic nonstick coatings are those composed of either fluoropolymers, such as polytetrafluoroethylene (PTFE) or of silicone polyesters. PTFE-based coatings have been widely used on cookware, bakeware, and small appliances. Middle-market PTFE coatings are two ply with a binder-containing base coat and a fluoropolymer-rich finish coat. The most durable PTFE coatings for high-end cookware are three-coat systems with a ceramic oxide-containing intermediate coat for scratch resistance.[5] Silicone polyesters are widely used on the exterior of cookware in a variety of colors. These two families of organic materials have a low surface energy in common, which prevents the adhesion of burned-on foods. Thus, for example, only water and paper towel can be used to remove residue with the application of minimal force. However, both types of coating are soft and easily scratched or gouged. The abrasion resistance of the PTFE coatings tends to be much less than porcelain enamel. Additionally, there are two health and environmental concerns with PTFE. The first involves a surfactant used in the manufacture of fluoropolymer resins, but this material is unlikely to be present in cured PTFE coatings.[6] The second is the emission of toxic by-products from PTFE during thermal decomposition, which can occur if, for example, cookware is overheated.[7]

RealEase™ marries the cleanability of the organic nonsticks to the durability of vitreous enamel. In a single coat, it offers the scratch resistance of enamel and the cleanability of PTFE. It is a patented technology that stands to bridge the gap between the organic nonsticks and low-temperature porcelain enamel.

Processing

RealEase™ is applied to degreased and roughened aluminum, aluminized steel, brass, or copper. Unlike conventional enamel, it can be applied to die-cast aluminum. Roughening can be done with sandblasting or acid etching. It can be applied to mild steel, stainless steel, cast iron, glass, or ceramics after the application of special patented hard bases. Hard bases consist of an enamel base coat with a rough surface.[8]

RealEase™ is supplied as a wet, ready-to-use (RTU) system. The RTU slip requires no adjustment of set, gravity, or color. Like enamels and unlike PTFE, the overspray can be reclaimed and reused at 30%. The coating is dried at about 125°F (52°C). Once dried, it can be screen printed with other RealEase™ colors and fired in a single process. The coating fires at less than 1000°F (538°C). The time varies with the metal thickness and thermal conductivity, and a convection-type oven is preferred. Volatile emissions during curing are 70% to 80% less than PTFE.

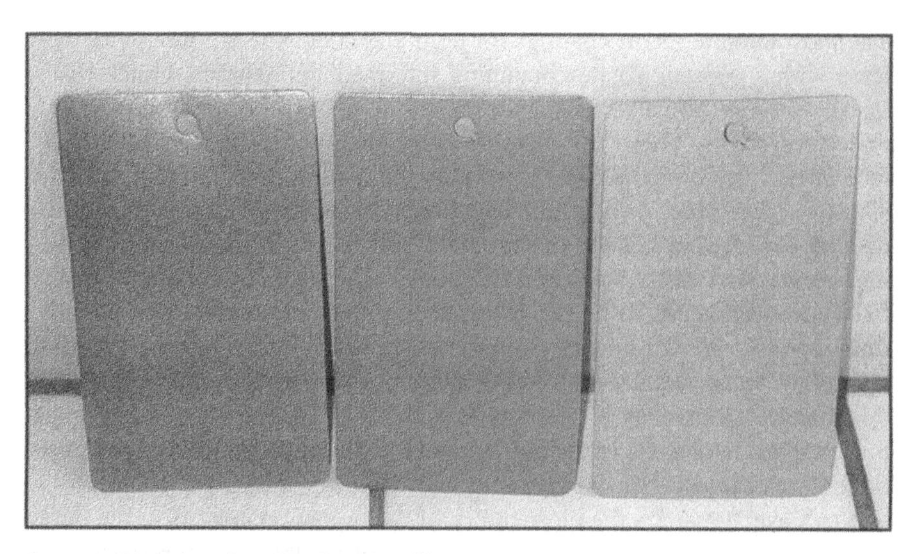

Figure 1. Examples of metallic RealEase™.

Fired Properties

Typically, RealEase™ is about 2 mils (50–60 μm) thick without hard base and as low as 1.5 mils (40 μm) with hard base. The hard base is typically 1.5 mil (40 μm) thick and requires an additional fire.

RealEase™ can be supplied in many colors except bright white. Generally, the color palette is similar to that for aluminum enamels. RealEase™ has a satin finish with a 60° gloss of 2–10, depending on firing conditions. Possible effects are mottles, stipples, shadow application, and recently developed metallic colors. Examples of metallic RealEase™ colors are shown in Figure 1.

RealEase™ has lower surface energy than conventional enamel and, therefore, is hydrophobic and easy to clean. The balance of forces arising from a sessile drop of liquid water (*l*) on the coating surface (*s*) under a vapor (*v*) is schematically shown in Figure 2 and is described with Young's equation,

$$\gamma_{s/v} = \gamma_{s/l} + \gamma_{l/v} \cos \theta$$

where γ is the energy per unit area of the appropriate interface and θ the contact angle between the liquid and the substrate. If $\gamma s/v > \gamma s/l$, the surface will be wetted to decrease the area of the higher-energy s/v interface; this is the situation with conventional enamel. If $\gamma s/v < \gamma s/l$, balling up of the water will occur to reduce the area of the higher-energy *s/l* interface.

Typically, θ is only 60° for conventional enamel, about 120° for PTFE, and about 110° for RealEase™. This is shown in Figure 3. Droplets of

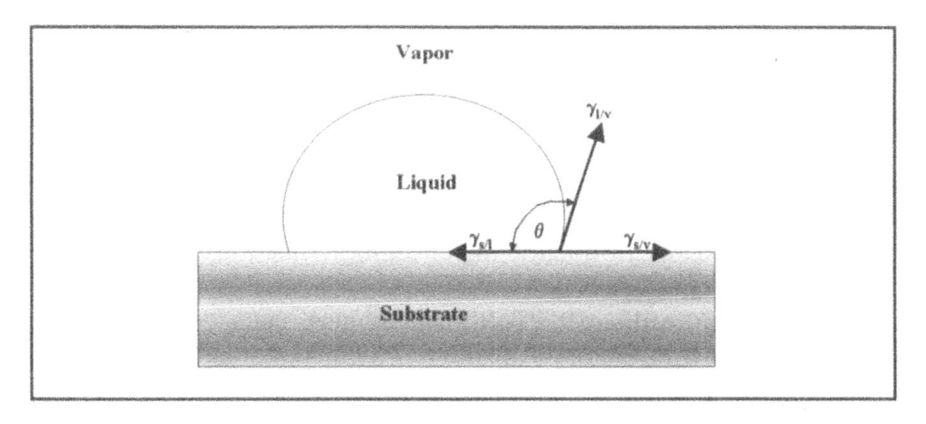

Figure 2. Liquid droplet on a hydrophobic surface.9

Table I. Easy-to-Clean Test-Scoring System

Step	Cleaning method	Score
1	Surface of coupon completely cleaned by wiping with a dry cloth.	5
2	Surface of coupon completely cleaned by wiping with the soft side of a sponge and soaking solution.	4
3	Surface of coupon completely cleaned by wiping with the abrasive side of a sponge and soaking solution.	3
4	Surface of coupon completely cleaned by wiping with the soft side of a sponge and liquid abrasive cleaner.	2
5	Surface of coupon can be cleaned only, if at all, by wiping with the abrasive side of a sponge and liquid abrasive cleaner	1.

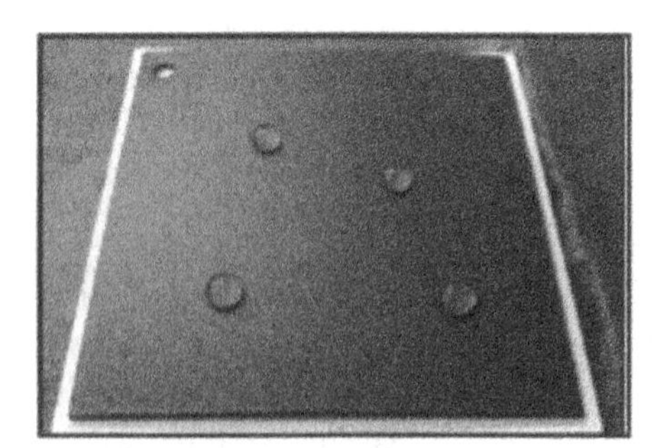

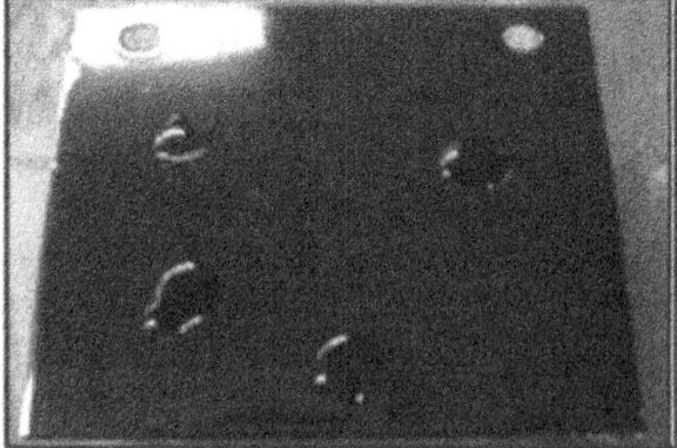

Figure 3. Wetting on (a) conventional enamel and (b) RealEase™.

water wet and spread on a dual-purpose black ground coat. On RealEase™, they are repelled by the surface and form beads.

The cleanability has been assessed by the European FAN (Facile à Nettoyer, literally, Easy-to-Clean) test. First, five steel rings are glued to the coating surface. Second, salted whole milk, gravy, lemon juice, egg yolk, and ketchup are placed in each ring. Third, the test panel is baked at 482°F (250°C) for 30 min. Then, the glue thermally decomposes, which allows the ring to be removed. Finally, the cleaning is rated using the system shown in Table I

The results for RealEase™, PTFE, stainless steel, and a typical porcelain enamel are shown in Figure 4. RealEase™ achieves a perfect score and shows cleanability equal to PTFE and much better than stainless steel or traditional enamel.

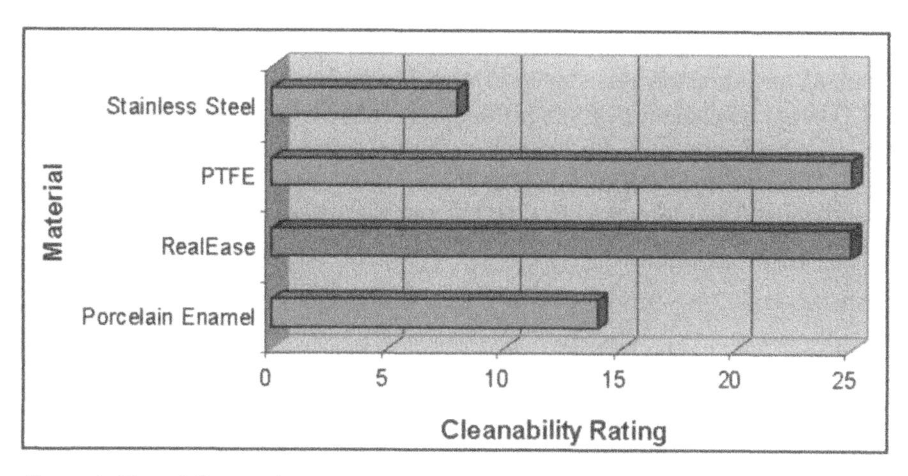

Figure 4. Cleanability results.

RealEase™ exhibits good acid and alkali resistance. It achieves a AA rating on the PEI T-21 citric acid spot resistance test. It is UV stable and unaffected by solvents. It has excellent scratch and abrasion resistance.

Scratch resistance is evaluated using ASTM D 3363-00 "Standard Test Method for Film Hardness by Pencil Test." The force required to gouge a coating with a drawing lead of calibrated hardness is assessed. RealEase™ shows a rating of 8H compared with an average of 4H for PTFE coatings. Results are shown in Figure 5.

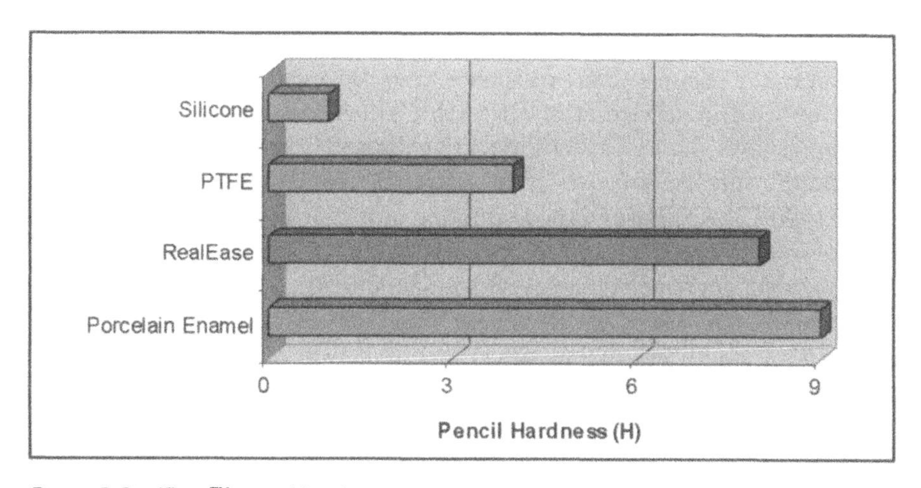

Figure 5. RealEase™ pencil hardness.

Abrasion resistance is measured using ASTM D 4060-95 "Standard Test Method for Abrasion Resistance of Organic Coatings by the Taber Abraser." The weight loss before and after 2000 cycles under the most abrasive CS-17 wheels under a 1 kg load was measured and normalized to the coating thickness. Results are shown in Figure 6. On average, RealEase™ was less damaged by abrasion than PTFE and significantly less than silicone polyesters.

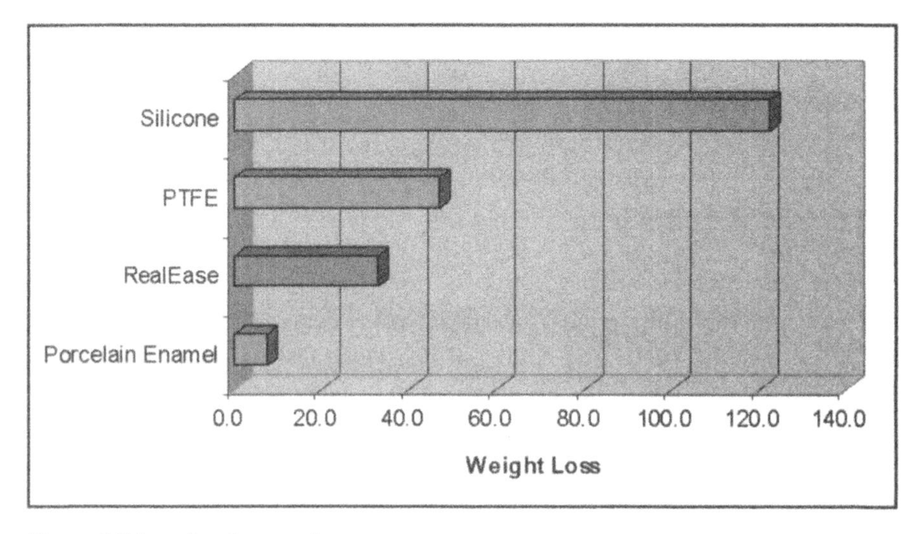

Figure 6. Taber abrasion results.

RealEase™ has more heat resistance than the organic nonstick coatings. It can be used in service at 600°F (315°C) for extended periods of time. Figure 7 shows the color stability of RealEase™ compared with a high-temperature silicone polyester paint. After 100 h at either 662°F (350°C) or 752°F (400°C), RealEase™ showed little color change versus a ΔE as high as 51.17 for the silicone polyester after 100 h at 752°F (400°C). PTFE shows similar degradation at elevated temperatures.

Potential applications for RealEase™ include the interior or exterior of all types of cookware, including ceramic bakeware. RealEase™ is listed with the National Sanitation Foundation (NSF) as compliant with Standard 51. It is certified as safe for food contact (acidic, aqueous, <8% alcohol beverages, dairy, dry solids, and other) up to 750°F.[10] With superior heat

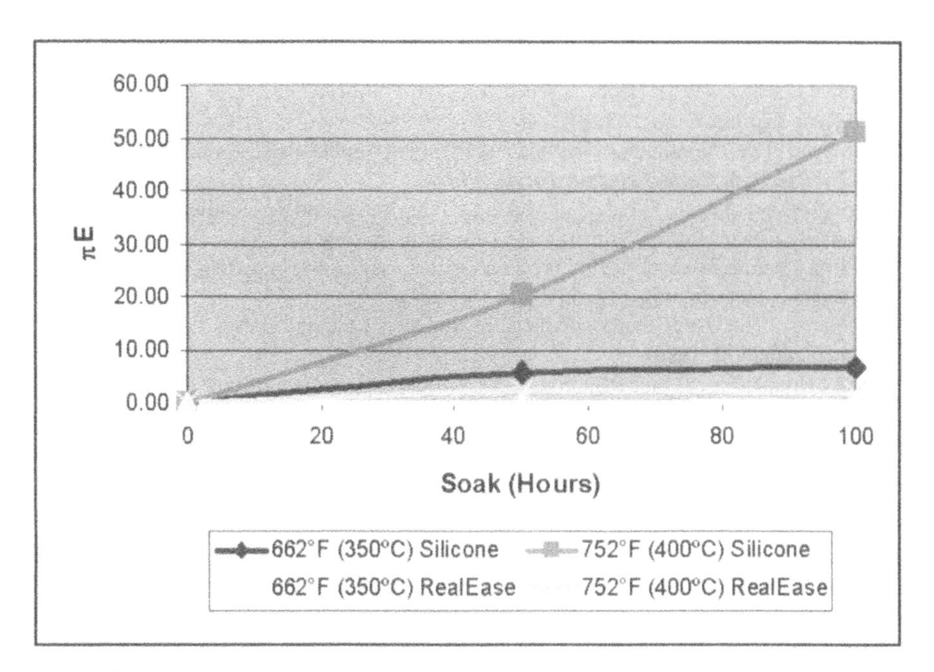

Figure 7. Color stability of RealEase™ versus silicone paint.

resistance, it is very promising for heavy-duty use environments, such as commercial kitchen restaurants. It can be used on small appliances, such as toaster ovens and microwaves. It is an alternative to silicone polyesters and PTFE on grills, griddles, and simmer plates.

In summary, RealEase™ is the first true nonstick porcelain enamel. It is water-based and applied as a single coat. Compared with silicone polyesters and PTFE, it is much more scratch resistant, more abrasion resistant, and more heat resistant.

Acknowledgements

RealEase™ was discovered by Alain Aronica, Remy Caisse, and David Coutouly. Thanks to Nael Amir, Ayman Nouman, Dave Fedak, Sebastien Bonneau, Patricia Secrest, and the Ferro CGPM Lotus project, Lou Gazo, Holger Evele, the rest of the Porcelain Enamel Development Lab, and Darry Faust for their contributions. Thanks to Renee Pershinsky for proofreading of this paper.

References

1. William D. Faust, "What Happens When You Cook?" *Proceedings of the 53rd Porcelian Enamel Institute Forum,* 1991.
2. W. Stephen Lee, "Continuous Cleaning and Pyrolytic Applications," *Proceedings of the 31rd Porcelian Enamel Institute Forum,* 1969.
3. H. Berkenkoetter et al., "Method of Making a Temperature and Scratch Resistant Anti-sticking Coating," U.S. Pat. No. 3 373 290, April 16, 2002.
4. "U-Jin's Coating System," U-Jin Porcelain Enamel, Ltd., 2001, http:www.ujin-enamel.co.kr/technology_e.html (April 12, 2004).
5. P. Thomas, "The Use of Fluoropolymers for Non-stick Coating Utensils," Surface Coating International 12, 1998.
6. Cheryl Hogue, "Focus on Perfluoros," C&EN Today's Headlines 2003. http//pubs.acs.org/email/cen/html/042103120125.html (April 21, 2003).
7. David A. Ellis *et al.*, "Thermolysis of Fluoropolymers as a Potential Source of Halogenated Organic Acids in the Environment," *Nature (London),* **412** (2001).
8. William D. Faust, "Ceramic Substrate for Non-Stick Coatimgs," *Proceedings of the 63rd Porcelian Enamel Institute Forum,* 2001.
9. Charles Baldwin, "Microstructural Engineering of Alumina- and Zirconia-Based Laminates Using Rapid Prototyping"; M.S. Thesis. Case Western Reserve University, 1999.
10. "NSF Certified Products—Food Equipment," *NSF International 2004.* http://www.nsf.org/Certified/Food/Listings.asp?TradeName=Real+Ease&CompanyName=&PlantState=&PlantCountry=&PlantRegion=&Standard=051&search=SEARCH (April 19, 2004).

Aluminum Stove Grates—An Innovation in Cooking

Dave Thomas
American Trim, Erie, Pennsylvania

Abstract

Aluminum stove grates were introduced as a potential alternative to cast iron grates during the 63rd PEI Technical Forum.[1] Thermal imaging was used to demonstrate the temperature profile differences between cast iron, steel-wire, and cast aluminum grates. Aluminum grates were shown to exhibit more uniform heat distribution and lower operating temperatures during testing.

This presentation demonstrates the suitability of cast aluminum grates for use based on industry standard tests including the American Gas Association (AGA) brick, thermal shock, flame impingement, and time-to-cook tests. No structural damage was observed on the AGA brick load test, as shown in the accompanying illustration. Time to bring a prescribed quantity of water to boiling showed a faster time for cast aluminum versus cast iron. Additionally, the cast aluminum grates cooled more quickly than the cast iron grates. Is was calculated that cast aluminum grates also will reduce shipping costs by approximately 31% because of their lighter weight for comparable cross sections.

- The benefits of aluminum grates are
- No structural damage on AGA brick test;
- Superior thermal shock resistance;
- More efficient heat transfer;
- Hotter cooking surface, faster heat up, and faster cool down;
- Excellent flame impingement properties;
- Potentially lower shipping costs;
- No rust or etching from cleaning grates in a dishwasher; and
- Availability of a variety of colors and coatings not feasible with cast iron

Some new nonstick coatings, such as RealEase®, supplied by Ferro Corporation, may be applied to aluminum grates to produce a nonstick surface.

These coatings have sufficient high-temperature performance characteristics that would allow them to be used on grates. The nonstick surfaces provide easy-cleaning options.

Reference

1. Charles A. Baldwin, "Thermal Imaging of Enameled Aluminum Pan Supports," *Proceedings of the 63rd Porcelain Enamel Institute Technical Forum, Ceram. Eng. Sci. Proc.,* **22** [3] 97–105 (2001).

Porcelain Enameling of Cast Iron via Electrophoresis

Julie Rutkowski
Roesch, Inc., Belleville, Illinois

Introduction

Roesch has been enameling cast iron for more than a decade, during which time its methods and quality have changed dramatically. Initially, Roesch used traditional wet-spray applications to coat cast iron, and now it has progressed into the world of electrophoresis. This paper will focus on the electrophoresis process of porcelain-enameling (EPE) cast iron and the obstacles encountered along the way.

EPE Process General Overview

The grates are loaded onto the EPE carousel, at which point the multiple-stage enameling process begins.

1. The first stage in the EPE process is a contact-cleaning bath. The electrical contacts are submersed into a tank of caustic soda, where a rectifier delivers a charge to the rack system within the tank and the carousel tooling. This step is very important, because it will affect the transfer of electrical current to the part.
2. Next, the grate is rinsed to remove any cleaning solution before it enters the copper activation bath.
3. During this stage, the part receives a thin layer of copper to add additional electrical conductivity, which aids in the enameling process.
4. The part is once again rinsed via dip and spray methods before it enters the enameling tank.
5. The EPE rack is submersed into the enamel tank, where it is agitated and receives an electrical charge from the rectifier. The charge is applied to the enamel material and the cast-iron pieces.
6. After the enamel material has been applied, the part receives a dip and spray rinse to remove excess enamel.

7. The part is subjected to air knives to aid in removing water droplets, and then it is processed through a convection dryer.
8. Enamel touchups are made, and the part is transferred to the furnace chain for the firing process.

Enamel Parameters

Roesch monitors several parameters on the EPE system. The EPE system is more complicated than conventional wet spray systems and requires constant attention of specific gravity, viscosity, pH, conductivity, salts, temperature, and wrap around.

The enamel must be of the proper specific gravity, viscosity, pH, conductivity, salt concentration, and temperature, and it must exhibit good wraparound characteristics. It should be noted that maintaining control of the aforementioned parameters is critical to the EPE coating process; therefore, these parameters must be monitored and recorded on an hourly basis during the production run. If the temperature and conductivity increase beyond the preferred range, the enamel will bubble during the EPE process and cause bisque and fired defects. Low salt levels allow for oxygen to collect at the work piece, thus creating a porous coating. Each EPE system is different, and each enamel is unique; therefore, Roesch has developed operating target ranges for these parameters.

Enamel Formulations

- EPE enamel formulations have limitations and must take into account the following conditions:
- Preferential deposition;
- Particle-size distribution;
- Oxide floatation; and
- Oxide loading.

EPE enamels have reduced milling times and oxide additions. The long milling times generate more heat and increased enamel conductivity. Freshly milled material should have a lower conductivity and pH, which allows for substantial chemical adds. Metal oxide additions in excess of 2.5% rarely contribute to the color of the final product. When choosing oxides, keep in mind the oxide density, because this can lead to floatation within the EPE tank. During the EPE enameling process, only a certain amount of particles are deposited regardless of the pigment or frit content. EPE enamels have preferential particle deposition. One frit may deposit more quickly than another frit or an oxide. Past experience has shown that it is best to have the majority of the pigment smelted into the frit and keep the frit formulation to a minimum. Keep in mind those subsequent production millings often times differ from initial enamel formulations. Unlike traditional spray enamels, EPE enamels might need clay concentration adjustments because of binder break down over time or increased fineness because of particle-size distribution.

EPE Process Obstacles

Electrical field lines must be considered when determining part orientation in relation to the electrodes. Originally, the EPE design contained a center electrode; however, after evaluating the results of a number of trials, it was decided to eliminate the center electrode and reduce the combating fields. Even though the part is totally submersed in enamel, it does not guarantee that the part will be completely coated. Enamel moving in one direction can prevent deposition at the part by conflicting with enamel moving in the opposite direction. The wrap around or the voltage drop ratio is a good coating indicator. Depending on the loading of the rack, the higher the wrap around value, the more uniform and complete is the porcelain enamel coverage.

After working through the tooling configuration and enamel hiccups, a white milky film appeared on the parts before and after firing. Roesch decided to look at the following variables:

- Rinsing
- Enamel formulations
- Weather conditions
- Enamel viscosity and
- Firing conditions.

The initial thought was that something had changed in the rinsing. After reengineering the EPE rinse features, the problem was still present. A series of tests were performed. Enamel formulations were analyzed, processing steps were reviewed, and weather conditions were documented. Finally, it was determined that firing the casting at a higher temperature eliminated the film.

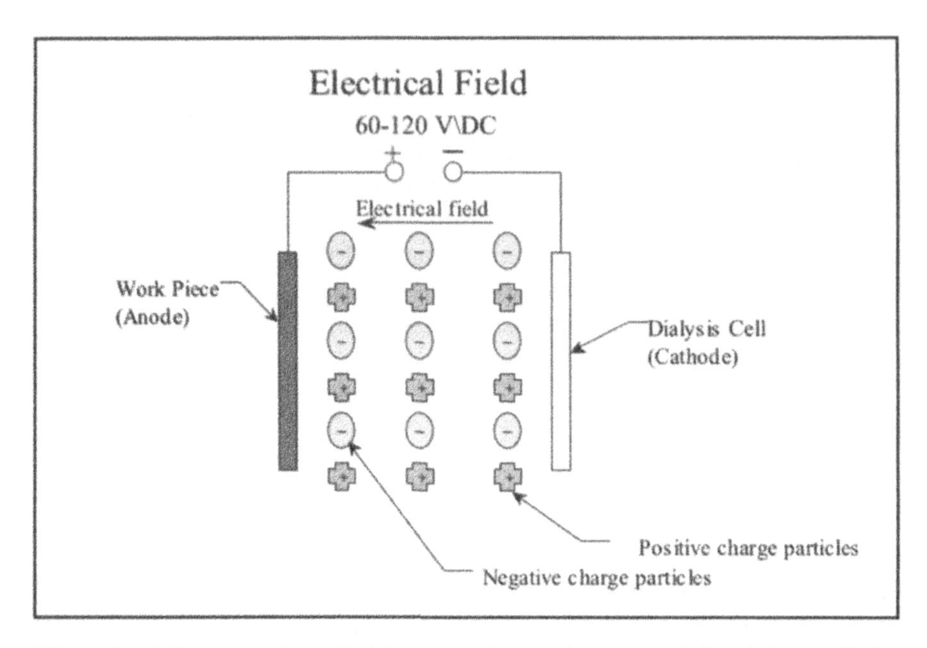

When the DC current is applied between the work piece and the dialysis cell the porcelain particles become negatively charge by electro-chemical surface reactions and friction from their movement.

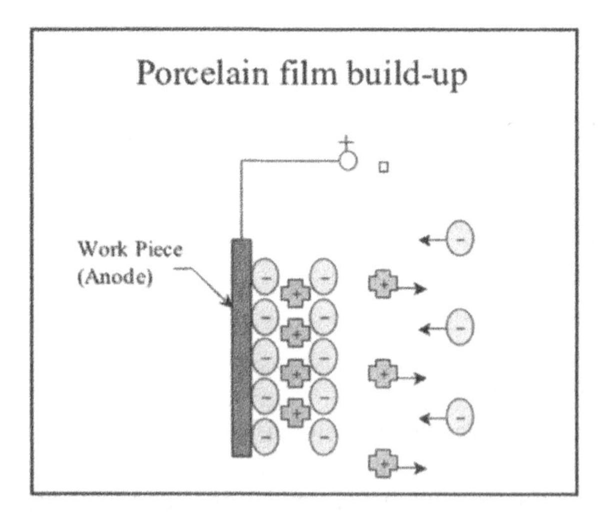

During the film build-up, Electro-osmotic dewatering of the bisque takes place because the water is positively charged. This results in a bisque with a water content of 22% (versus 50% for a regular wet bisque). the bisque is dense and very strong allowing manualtransferring to the furnace line without damaging the bisque.

After visiting EPE systems in Europe, Roesch equipped the laboratory with the necessities. The production machinery was scaled down, and then the EPE laboratory tank was designed to develop the closest correlation between testing and manufacturing. Roesch is able to evaluate sample enamels for application feasibility and enamel appearances. The laboratory-scale model is not accurate in predicting enamel performance or problems that will be encountered during the production process. Unfortunately, the real test is putting the enamel into the production tank and running numerous trials. Most initial start-up defects work themselves out over time with no process adjustments.

Benefits

EPE allowed Roesch to decrease the enamel usage per part while increasing enamel coverage, and enamel carried over into the rinse tanks can be 100% reclaimed. Electrophoresis allowed for fine-tuning of coverage and enamel thickness via parameter adjustments. More consistent enamel coverage decreases thermal shock and chipping failures. EPE gave Roesch the ability to provide appliance manufacturers with a high-quality completely enameled cast-iron grate, which is most desirable to the dishwasher society.

Summary

EPE has been challenging. As with any new system, EPE or powder, it is better to run, monitor the parameters, and wait before adjusting the variables involved.

Cast Iron Quality for Good Porcelain Enameled Parts

Liam O'Byrne
A B & I Foundry

Abstract

The physical and chemical requirements for cast iron enameling are important for successfully coated parts. Metallurgically, the cast iron has to have a variety of elements in relatively narrow ranges to meet application and processing requirements. The chemical composition of cast iron for enameling should have the following:

- Total carbon of 3.2–3.6 wt% (graphitic carbon of 2.8–3.2 wt% and combined carbon of 0.2–0.5 wt%);
- Silicon of 2.25–2.70 wt%;
- Manganese of 0.45–0.65 wt%;
- Phosphorus of 0.30–1.00 wt%; and
- Sulfur of 0.05–0.10 wt%.

Each element in the composition has critical tasks in making a successful product.

Carbon

Carbon exists in the metal matrix as primary graphite flakes of pure carbon or iron carbide (cementite, Fe_3C, chilled iron). The quantity of each phase depends on the overall composition and the cooling rate of the casting. Carbon promotes fluidity of the iron and increases the solidification rate. Graphite promotes excellent machinability and is the source of the "gray" color in the fractured surface of gray cast iron.

The graphite structure is heavily dependent on the cooling rate of the casting and interacts with silicon and phosphorus for optimal solidification of the graphite phase. The faster the cooling rate, the finer any primary graphite in the metal matrix and the higher the percentage of cementite in the matrix.

If too much carbon is present or the cooling rate is very slow, a "kish" graphite phase can solidify in the matrix. Kish graphite is detrimental to the enameling process and is almost impossible to remove from the surface of the casting. Kish graphite is very rarely seen today in enamel iron foundries as a result of the increase in use of electric melting furnaces.

Too little carbon promotes chilled (white) iron, especially near the cast surface. Almost all enameling castings are produced in green sand molds, where water is a key component of the mold formulation. Rapid cooling of the metal at the mold surface by the moisture in the mold usually results in a very thin layer of chilled iron at the surface of most solidified castings. This needs to be removed for successful enamel processing of the castings.

Silicon

Silicon promotes primary graphite separation during solidification. It is normally added at the melting stage or as an inoculant during the pouring stage of casting. Silicon improves the fluidity of the molten iron and increases the solidification range. Too much silicon causes coarse primary graphite with associated outgassing defects and can cause shrinkage in heavy cross sections. Too little silicon will promote a chilled iron structure.

Phosphorus

Phosphorus greatly affects the fluidity of the iron, with increasing amounts increasing the fluidity. It also increases the solidification range of the iron, lowers the pouring temperature, and assists with the formation of the graphitic structure. The iron–phosphorus eutectic (steadite) is generally the last to solidify in the grain boundaries of a casting. Large quantities ($\geq$0.5%) of phosphorus can produce an almost continuous network of steadite around grain boundaries that can make the iron harder and more brittle.

Carbon Equivalent

Carbon, silicon, and phosphorus react to produce a value called the carbon equivalent (CE):

$$\text{CE} = \%\ \text{total carbon} + (\%\ \text{silicon} + \%\ \text{phosphorus})/3$$

The ideal range for the carbon equivalent is 4.2–4.6%. Thicker cross sections can be poured with lower CE values and thinner sections can be poured with higher CE values.

Manganese and Sulfur

Manganese must be present in sufficient quantity to neutralize the effect of sulfur in the cast iron formulation. The usual ratio is

$$\% \text{ manganese} = 1.7 \times \% \text{ sulfur} + 0.3\%$$

Sulfur is a strong carbide former and also causes porosity and brittleness of the iron. These contribute to severe outgassing of the enameled product. Manganese combines readily with sulfur to form manganese sulfide (MnS), which does not have these detrimental effects. If the sulfur is not properly balanced, excess sulfur causes formation of a very fluid slag with the iron that is extremely difficult to remove. The resulting slag defects cause outgassing and blistering defects in enameled articles. Excess manganese fully binds with sulfur, which prevents slag formation and toughens the iron.

Although each of these chemical components in the casting significantly contributes to the final properties of the casting and the likelihood of success in the enameling process, the chemical composition is only the first of a number of variables that must be adequately controlled for consistency in enameling cast iron. Future presentations will address these issues in detail.

Robotics in the Job Shop

Randy Smitley and Jeremy Foster

Pryor to June, 2002, the Porcelain Metals Corporation (PMC) wet spray process consisted of automatic electrostatic guns and wet reinforce. This provided PMC customers with an acceptable cosmetic appearance on all the parts we ran through this system. The problem for PMC was the same as everyone in our industry faces: a highly labor-intensive process. Even the flatware we ran through the automatic electrostatic system was labor intensive. As for complex shapes such as high end oven cans or boxes we could not employ enough sprayers to make good parts consistently. Add to this the labor shortage we experienced in Louisville in the late 1990s through 2002 and you have a situation that was unbearable for us and our customers. Spraying had become the least desirable job at PMC. We actually ended up in the position where we were forced to use untrained temporary labor on several of the spray jobs. As we all could guess, what followed was a period of time where we had burnoff on one part and sags and runs on the next part.

We began to look for an alternative to hand spraying. Robotics was the natural choice for our particular product mix. We knew that several of the major appliance manufacturers were using robotics; therefore, we called a few suppliers in to see what was offered. We knew going in that the biggest objectives would be to reduce labor, improve quality, save material, and gain consistency in the process. We knew that these objectives would hinge on the dependability of the unit we chose and the ease of programming and maintenance of this unit.

It was about this time that Jeremy Foster contacted us. Foster represents a line of robots from Artomation (Cleveland, Ohio). After looking at other presentation of robots that operated from a fixed position, the Artomation units were breaths of fresh air. Their units operate on a gantry, which allows them to move with the conveyor line. This was important to us because we wanted to do as much as possible with one unit. The unit let us approach the programming as if we were instructing an individual to hand spray.

After viewing the Artomation presentation, we made numerous trips to their lab in Cleveland. We hauled examples of all the different parts we were running at that time to their lab to prove or disprove that this was indeed a feasible project. In the beginning, the Artomation engineers were working with wet paint, because they had no experience with porcelain enamel. They developed some basic machine movement programs; e.g., a movement that could coat a piece of flatware. After that relatively simple task, they moved to oven cans. After much trial and error, they were successful in the development of a "box Program" that would allow the spray gun and robot arm to articulate inside an oven can. After this was developed and tried with paint in their facility, they called us in to approve or disapprove.

As I stated earlier, the Artomation people had absolutely no experience with porcelain enamel. All the timing and defaults built into their programs had to be modified to allow for the heavier material build required by porcelain. Many cans were sprayed and washed off by hand to prove that we could do this. As with training a person to spray, so it was with the robot: thin spray and sags. We discovered that the best way to regulate film build in certain areas of various parts was by changing gun speeds. We were running off a pressure tank; therefore, instantaneous pressure changes were not possible. We were finally able to achieve this on flatware and cans.

Now that we could spray a stationary part, we had to introduce movement to the equation. Artomation has relatively simple line-tracking system setup in their lab. The flatware was not a big problem; however, the cans proved to be another story. Much banging of the inside of parts occurred, but persistence paid off. They now had a box program that could spray and track with the conveyor line.

A presentation was made to the owner of PMC, John McBride, and he made the decision to move forward with the project.

Once the decision to go ahead with the robot was made, we had to decide where to locate the unit on our line and what type of tooling to use. We had to decide how fast we could reasonably expect to run the production line.

We decided to locate our unit in the booth that housed the Devilbiss electrostatic equipment. This is the only enclosed wet booth we currently have at PMC and, therefore, the only booth that allows us to control humidity with steam and make-up air. We believed all along that one of the keys to making the robot successful would be to run in an atmosphere that would allow us to wet the enamel out rather easily. Once the decision on location was made, we had to raise the roof on the booth to accommodate the 12 ft. high gantry. We added 5 ft. of head space to the booth and knocked out one wall temporarily to allow us to gain entry into the booth with the robot. The unit was shipped completely assembled from Cleveland.

Some modification of the conveyor drive had to be made to accommodate the encoder for line tracking. We also had to install a control room for the computers that run the robot. There are two computers. One is used to program the various spray paths into the system. This same computer is used to start, stop, jog, and control the parameters of the spray gun. The other computer actually controls the motion system of the robot. It processes the information from the first computer and converts information into movement commands. These programs were written by the engineers at Artomation and are never accessed by PMC personnel.

After a 12 week lead time, the robot was completed. We made another trip to Cleveland for the run out at the manufacturer's facility. We experienced a few problems but no show stoppers. The unit was delivered and set up by Artomation personnel and our maintenance people. Once setup was complete, the Artomation engineer in charge of this project came to our facility for training sessions. One other person and myself from PMC had received some training at Artomation on this type unit. They have a smaller version of this unit set up in their lab, and they made this unit available to us about anytime we wanted. From the training we received in their lab along with the in-house training we received, we were able to train one of our hourly people to be the control person for this unit.

As I mentioned earlier, one of the reasons we went with this unit was ease of programming. We simply take a digital picture of the work piece we want to spray. We then transfer this image to the Artomation program and then draw spray paths with the computers mouse. After a path is drawn, we insert gun trigger points anywhere we want them. This is a simple mouse click procedure. After inserting the spray path and trigger points, the program asks us to fill in a few basic parameters: gun speed (in in./s), gun angles, width, length, and depth of the part. Artomation has inserted some basic tools into their program that make these tasks easy. These tools include lines, arcs, circles, and boxes. Using these tools, programming could not be easier.

Once all the requested parameters are entered into the program, we are ready to check gun movement. We first let the gun go through its motions without having a part in front of it. When we are satisfied that the movement meets our requirements, we center a part in the work frame for a dry run without spraying. This is where we can usually tell if we need to tweak they spray paths a bit. At this point, we can still add or delete a spray path or modify the direction of the path or the gun angles. The speed can be adjusted here if we believe it is necessary. After we have made all the adjustments we believe are needed, we hook up to a pressure tank and adjust our gun parameter (fan size, fluid pressure, and atomization) through the program. This is normally the quickest part of the setup or learning phase, because it does not change much from part to part. When we are satisfied with the setup, we add line movement. We tell the robot where to pick up the part and the direction of line movement. If we have done our job on all the other information we have given the robot, the addition of line movement does not effect the spraying of the part. We then adjust the line speed to the speed the robot needs to spray the parts. The line speed can be adjusted up or down, and the robot will track with it. However, we cannot exceed the speed of the gun movement or "drift" will take place. This is where the machine falls a little more behind with each part it sprays. Eventually the parts will be out of the work envelope before the robot can finish its spray paths. When this happens the machine shuts itself down.

Normal setup of a new part usually takes about one hour from the time we first see a part to being ready for production on the line. Once a set up is made and saved into the program, changing from one part to another when using the same enamel only takes about 30 s, or two empty hangers on the conveyor line. When we have to change enamel, the change over increases to about 10 min. i.e., time to wash out the enamel hose and gun and to change the pressure tank.

The beauty of this system is that it allows us to make adjustment on the fly. Spray paths or any other of the parameters can be changed while the robot is running production. As a matter of fact, we can install a completely different setup, save it, and, when we want to try it, send it to the robot and not miss a beat. If the change does not do what we wanted, we simply recall the previous program. We also have the ability to program on another computer in another location. I can program on my laptop, save to a disk and bring up the computer for the robot.

We can spray any of the parts we run on our robot complete. However, we have found that, in most cases, it makes more sense for us to use the robot as a reinforcer or touchup sprayer before we hand spray. The unit only has one gun, therefore, it is a little slow if the machine sprays a part complete. By using our method, we still enjoy all the advantages of the robot and still maintain our line speed. We have saved at least two hand sprayers on every job we run. In addition, we have saved 40%-60% on wet-enamel use over the old electrostatic system and/or hand spray. We have increased our through put, and yields are consistently in the 93%-97% range, even on hard-to-coat parts, such as oven cans. The biggest gain has been in the consistency of our spray process.

In an effort to constantly improve our process, we are experimenting with multiple gun mounts and multiple guns. We have tried several different gun types and sizes to improve the maneuverability of the robotic arm inside boxes. We also continue to look at the possibility of running two robots in sequence or the addition of a state of the art gun mover for spraying flatware.

In closing, it is plain that robotics is one way for job shop enamellers to take some portion of the variability out of the wet-spray process. Reducing labor costs and material savings are certainly enough to pique our interest, but the quality improvements we have experienced make this type of change a must for our survival.

Waste Minimization in Cleaning and Pretreating Operations

Charles G. Galeas Jr.
Technologies Group, Henkel Corporation, Madison Heights, Michigan

Pretreatment operations involving cleaners and conversion coating treatments generate wastes containing oils and greases, heavy metals, and high BOD/COD content. Several washer design techniques and modifications for minimizing these generated wastes will be discussed.

Introduction

Waste minimization for most pretreatment lines begins early during planning and designing phases. From the standpoint of planning a new coating line, the thought process is basically done as a reverse progression: first is the consideration of what are the desired performance requirements of the end product; then, back to the coating system requirements; followed by examining what pretreatment process will be necessary for successful coating; and finally to the challenge of pretreatment equipment design. After all the performance variables have been defined and pretreatment chemistry chosen to meet those specifications, the task of designing a cost- and quality-effective washer that will minimize the amount of generated wastes presents one with a wide array of options, much the same as buying an automobile today offers the individual consumer.

Existing pretreatment operations may have not utilized optimal design opportunities when the line was first installed, and operators may not be pleased with current operating costs, amount of waste being generated, or consistency in good coating development and performance. All is not lost for this group, because many of the design features, which are recommended for high end performance on new lines, can be adapted in existing operations.

The "Process Design" Process

After pretreatment design engineers have been provided with all the line performance requirements and have been introduced to the bath chemistries that will be used, what is their approach to designing a washer configuration?

In the real-world plant-working environment, one variable has typically been seen as the first limiting factor in pretreatment design: floor space constraints. It is not unusual for new projects to have preassigned a restrictive section of the plant for pretreatment construction, without even having preliminary investigations made into treatment requirements or washer design considerations. This is usually the one factor that will require design engineers to make "trade-offs" on optimum design characteristics.

For the purpose of this discussion, we are given the details of the pretreatment sequence recommended by chemical sales and technical experts for accomplishing the defined performance criteria of the overall coating system. A typical process sequence is illustrated in Table 1.

Table 1. A Typical Pretreatment Sequence

Stage No./Process	Contact time (min)	Operating temperature (°F)	Operating pressure (psi)	Comments
1/Alkaline cleaning	1.0–1.5.	130–160	20–50	Liquid product
2/Water rinse	0.5–1.0	100–150	20–50	Requires pre/post rinsing
3/Conditioning rinse	0.5–1.0	100–150	15–25	Slurry injection to header
4/Zinc phosphate	1.0–1.5	130–170	15–30	Three liquid products used
5/Water rinse	0.5–1.0	60–100	10–20	Requires pre/post rinsing
6/Chemical seal rinse	0.5–1.0	90–120	10–20	Non-chrome liquid
7/DI water rinse	0.25–0.5	60–100	10–20	Fresh and recirculation

Other variables collected during the design process will include:

- Work package geometry;
- Production rate and/or line speed;
- Required dwell time in each step;
- Temperature requirements;
- Application method—spray or immersion;
- Water quality requirements of each process step;
- Chemical properties of the pretreatment products; and
- Byproducts that will be generated from each process (oily waste, solids, sludge, heavy-metal wastes, etc.).

This brings us to the next and perhaps the most critical step of the design process: nozzle section and process solution exchange demands. Here, considerations are made in terms of each chemical process and the optimum mechanical properties for applying each solution. Alkaline cleaners and water rinses after cleaning generally require impingement and a significant volume of solution exchange, whereas other chemical processes are reactive in nature, such as phosphates and final chemical sealing rinses that require a low-impact "flooding" of solution over the substrate surface. In addition, consideration of byproducts formed in these chemical reactions may lead to the buildup of solid particles in the treatment solution, and, therefore, nozzle orifice size also becomes a key factor. Water rinses after phosphating need to be thorough; however, low impingement is necessary to ensure that no disruption of the newly formed crystalline coating takes place. Finally, nozzles and their associated spray risers always must be designed to anticipate the future maintenance requirements of cleaning and replacing nozzles that eventually will be needed. A typical nozzle arrangement for pretreatment operation suitable for most coating processes is shown in Table II.

Table II. General Recommended Nozzle Selection for Each Stage

Stage No./Process	Preferred nozzle type	Other nozzle options	Desired spray properties
1/Alkaline cleaning	Flat V	None	High volume/pressure, impingement
2/Water rinse	Flat V	K type	High volume/pressure, no plugging
3/Conditioning rinse	Flat V	K type	High volume, flooding effect
4/Zinc phosphate	Hollow cone	K type	High volume, low pressure, flooding
5/Water rinse	Flat V	K type	High volume, low pressure, no plugging
6/Chemical seal rinse	Full cone	K type	High volume, directed spray, low pressure
7/DI water rinse	K type	Full cone	Soft flooding, high volume, low pressure

At this point in the design process, the design engineer needs to evaluate the production rate, geometry of the work piece (either the single part or a single rack of parts), and line speed variables that are key to the entire plant production process. When these data are compared with the required processing time of each pretreatment stage, the calculation of spray zone length (or immersion tank length, for those processes) is quite straightforward. However, it is important to note here that, when pretreatment specialists speak of solution contact time, they are generally referring to the time in which the work piece is first entirely covered with the processing solution until the last point of spray contact. Therefore, in practical terms, design engineers will consider contact time to be between the first and last riser sets of the spraying system in a stage and not the length from the entry to exit ends of the spray area silhouette.

In addition, consideration is made with regards to the geometric shape of the work piece, or hanging pattern of a rack of parts, and the spray vestibule size requirements. Riser spacing is then calculated such that it will be uniform (typically 12, 15, 18, and usually no more than 36 in.) and generally 5–7 s apart, based on conveyor speed and utilizing a spray density formula (see Figures 1 and 2). In addition to these general rules, there are special considerations to be made for reactive treatments; for example, a zinc phosphate coating develops most rapidly during the first portion of the treatment cycle, and risers spacing is varied to flood more solution on the work piece at the entry side of the vestibule, with a greater riser spacing in the last half.

Having reached the point of laying out riser configurations, total nozzle requirements are calculated based on nozzle spacing that will provide the adequate coverage of the work piece: where no unusual spray conditions exist, nozzles are typically chosen so that spacing of 8–12 in. at a 40–50° spray angle satisfies coverage requirements. When the spray characteristics of each nozzle type are known, total solution flow rate requirements can be then

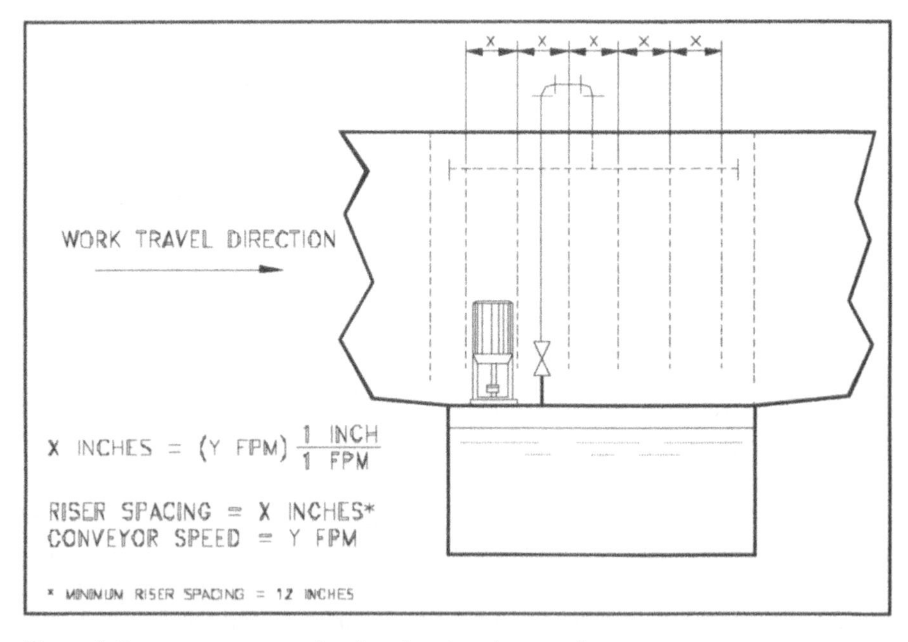

Figure 1. Pretreatment stage drawing showing riser spacing.

calculated for the desired spray pressure of the stage and process in question. Calculation of the total flow requirements lends itself then to the determination of pump size, with the final pump recommendation made to have at least 10–25% spare head capacity. Finally, tank sizes can be determined: typically, design engineers will specify that tanks for all chemical stages should be at least 3 times the pump delivery rate per minute and that all other tanks be 2.5–3 times the pump rate, but no smaller than 500 U.S. gallons.

Once tank sizes have been set, heat requirements need to be determined, based on the workload and as the pretreatment chemistry dictates. To determine BTU requirements, standard formulas and heat curves familiar to all design engineers are used. Once these calculations are made, heating methods are reviewed and chosen based on efficiencies unique to each process. In general, steam-plate-type coils, pipe coils, or external plate and frame heat exchangers are primarily recommended here, with gas-fired tubes as a lower-operating-cost alternative, where steam systems are not available. Gas-fired tubes do have lower demands on floor space, but they may consume washer chemicals and generate excessive sludge waste if used in these stages, and, generally, they are not recommended for these process tanks.

At this point, all data have been gathered and all calculations made, such that the first pretreatment washer schedule can be prepared, as shown in Table III.

There are other considerations made here in the review of nozzle selection that can significantly impact initial capital requirements and impact the amount of wastes that will be generated by the pretreatment line. All of the processes listed in Table III are accomplished through the combination of

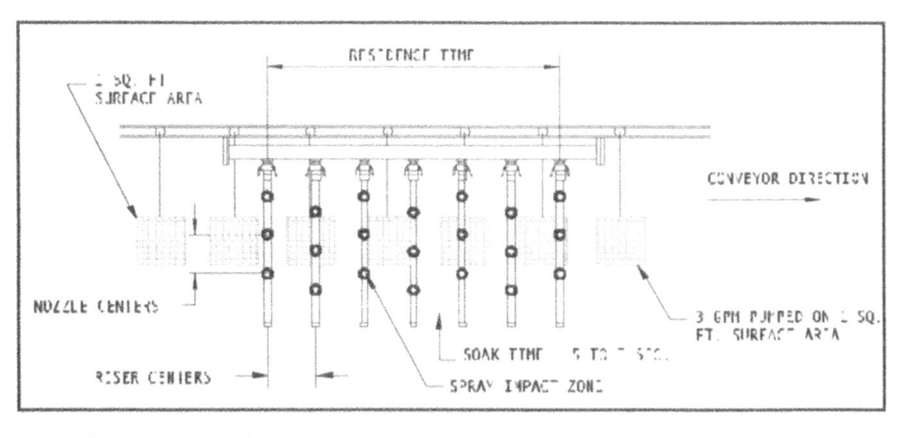

Figure 2. An example of a spray density calculation.

Table III. A Typical Pretreatment Washer Schedule

WASHER SCHEDULE

ZONE	TIME CYCLE	OPERATING TEMP.	NOZZLES NO.	TYPE	PRESS	GPM/EA	PUMP HEAD PRESS	GPM	SOL TANK CAP GALS	MOTOR HP	RPM
CLEANER	90 sec.	130-170°F	254	1.50" HST-5070	20PSI	4.9	55'	1150	3700	25	1750
RINSE	30 sec.	100-150°F	94	1.50" HST-5070	20PSI	4.9	55'	475	1400	15	1750
RINSE	30 sec.	100-150°F	94	1.50" HST-5070	20PSI	4.9	55'	475	1400	15	1750
PHOSPHATE	60 sec.	130-170°F	160	1.50" HST-50	15PSI	6.1	45'	990	2900	20	1750
			10	1.50" HST-9560	15PSI	3.7					
RINSE	30 sec.	NO HEAT	94	1.50" HST-5070	15PSI	4.2	45'	425	1200	10	1750
SEALER RINSE	30 sec.	NO HEAT	70	1.50" HST-9560	10PSI	3.0	35'	225	700	5	1750

time, temperature, volume, and concentration. Selection of appropriate nozzles and the proper spacing can result in fewer nozzles, smaller heating system required, and reduction of the potential for cross-contamination of the various processes and, hence, reduced generated wastes.

The design engineer's only remaining task in the basic design process of a washer is the completion of the floor layout for the entire process. In the connection of the various stages to each other, the task is to design the drain zones before and after each vestibule, within the limitations given on floor space. This is a key step in minimizing generated wastes: parts that have sufficient time to allow excess solution to drain back to the tank where it came from will reduce the need to dilute contamination drag-through to the next stage. The length of the drain zone is again measured as the distance from the last riser in one stage to the first riser in the following stage, and it should be as short as practical but of sufficient length to allow for proper drainage from the work piece, while preventing overspray to the adjacent stage. Typical drain zones should be in the range of 30–60 s; longer drain areas often will allow the work being processed to dry, or partially dry, which is undesirable from a quality standpoint. Drying of work can be prevented by the installation of wetting risers using the process solution of that particular stage, or the installation of misting nozzles. Drain boards are installed, pitched at a rate of 0.25–0.50 in./(ft of the drain zone length), to return overspray and drippings from the work piece to the proper stage. The apex of the drain board is usually located approximately two-thirds of the drain length to return drainage from the work piece to the preceding stage.

Finally, the design engineer has finished the task of preparing the basic pretreatment layout (see Figure 3) and completing the process design process and is ready to investigate means of process optimization.

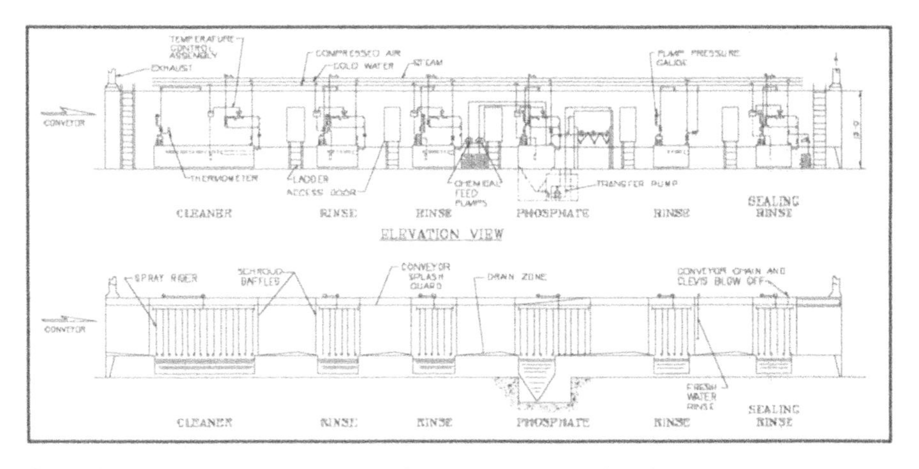

Figure 3. A basic pretreatment layout for a six-stage zinc phosphating system.

The "Process Optimization" Process

It is at this point that the process can be examined for all opportunities to enhance the pretreatment quality and economics of the washer's performance, while considering all options for waste minimization, during this process optimization. However, all of the techniques to be described here, although most easily accommodated during the initial period of design and construction, are readily adaptable to most existing lines that currently do not have these features.

Use of Level Controls

One of the greatest concerns in the design of a pretreatment process is the amount of fresh water that is required or consumed and the volume of effluent discharged from the system that will later become the burden of the waste treatment system. The use of level controls in chemical process and water rinse tanks is critical in managing a water use program. Level controls are available in many different styles, from simple float and valve actuation devices, through "bubbler" systems, to more expensive techniques, such as ultrasonic and radar level measurements. Levels are generally adjusted in water rinse tanks to maintain the operating level just at the overflow trough of the washer, where floating oils can be removed. Chemical process tanks need to be maintained below the overflow trough; therefore, when the circulating spray pumps are turned off, the volume of solution held in the spray headers and risers can be returned to the tanks without creating an overflow.

Many newer systems are originally equipped with some form of level control for all tanks, although these systems are not always adequate and may require upgrading in the future for more precise maintenance of tank levels. Level control systems remain one of the easiest and more cost-effective means of upgrading older washers: tanks can normally be fitted with a level control system inexpensively, with water savings and waste treatment minimization providing a payback period of no greater than 2 years.

Use of Counterflows

Further water conservation and waste minimization can be achieved when a level control system is utilized within the scheme of a well-designed counterflow system. Counterflow is primarily achieved with the movement of rinse water from later stages back to similar rinse tanks appearing earlier in the process sequence. The principle here is that a second or third water rinse after a chemical process is usually cleaner (less contaminated) than the earlier rinse stages. However, contaminated rinse water in a third rinse stage would be considered as "good quality" make-up water for the second and first rinse stages and, therefore, does not necessarily need to be discharged to the waste collection system once its maximum level of contamination is reached.

The installation of a counterflow system, prepared initially for new lines or easily added as an existing line modification, is accomplished with piping that is connected to the tank's spray header and solenoid valves installed in-line, permitting or preventing the backflow of the rinse solution. Solenoid valves are actuated by "on/off" timers, solid-state or programmable timers, or, in more sophisticated systems, automatic measurements of the rinse water conductivity and the maintenance of the rinse water below a maximum conductivity set point. Conductivity limits, or maximum contamination levels of each rinse stage, are normally set by the pretreatment chemical specialists and are usually represented as a maximum percent contamination of the chemical process bath solution they follow. A typical configuration for a three-stage rinsing process utilizing a counterflow system is illustrated in Figure 4.

A counterflow system is easy to install and brings significant cost savings to new and existing washers, again in the areas of water use and waste generated. With solenoid valves, piping, timers, and/or wiring to a chemical controller, capital investment is low and payback of 1–2 years can be expected.

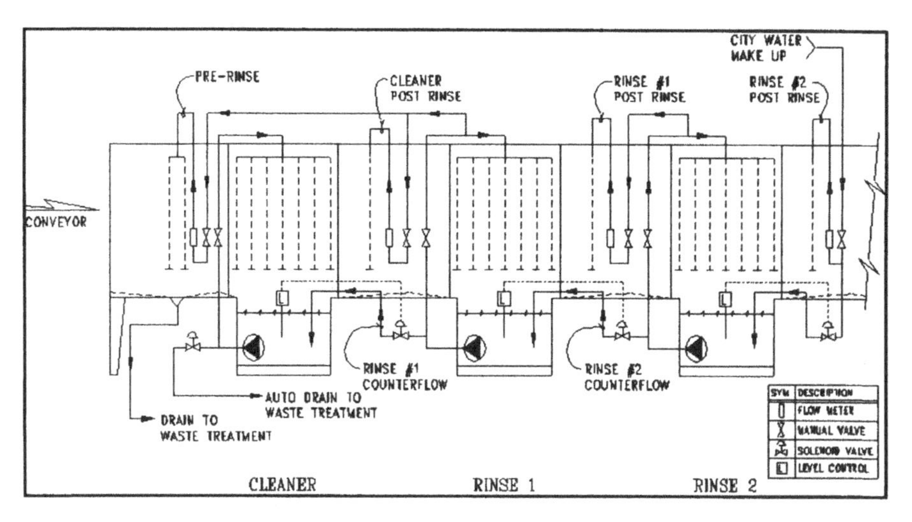

Figure 4. A typical counterflow system for rinse stages in a pretreatment washer.

Use of Prerinse and Postrinse Risers

Although the previous two methods of process optimization have been targeting waste minimization, the installation of prerinse and postrinse risers is an inexpensive means of improving the quality of the pretreatment operations and enhancing the final coating appearance.

One of the greatest concerns of the pretreatment specialist is the drying of the work piece between stages. Chemical solution and contaminated rinse solutions allowed to dry on the metal surface may form corrosive, reactive salts, which may not be possible to remove in the remaining washer stages. These salts left on the exiting surface will have a coating applied on top and later become sites for the initiation of corrosion. Dried solution also can leave a "streaking" appearance on the work piece, which may become apparent when closely examining the coated surface. With these potential failures, the use of prerinse and postrinse risers can keep all salts soluble and in a state that best facilitates thorough removal.

Again, this technique has been applied to new and existing lines: process or rinse solution from the stage in which the work piece is entering is plumbed to the prerinse and postrinse risers, or, alternatively, the postrinse risers plumbed from the subsequent process stage, and nozzle design is chosen to minimize volume requirements (and additional pump capacity requirements), while maintaining a "misting" environment for entry and exit travel of conveyor-moving work pieces (see Figure 5). Capital investment for this enhancement is low, with payback difficult to quantify, but, if viewed as an "insurance policy" against future failures or claims, the premium is quite reasonable.

Use of Process Control Equipment/ Automatic Chemical Feed Equipment

The use of process control equipment plays a key role in reducing waste of pretreatment chemicals that will be consumed as well as the amount of fresh water added to the system. Excessive chemical consumption usually means that only a portion of the chemicals being added are performing the desired surface reaction on the work piece surface, while another significant portion is generating undesirable sludge byproducts and additional waste that requires management.

Great advances have been made over the past 10 years in the area of process control equipment for pretreatment washers, in terms of capabilities and with regards to capital investment requirements. Washer process control equipment built as late as the 1980s had electrical panels 30 in. by 30 ft in size; today's control systems can be 30 in. by 30 in. in size, with 5–10 times the capabilities. From solid-state systems, through programma-

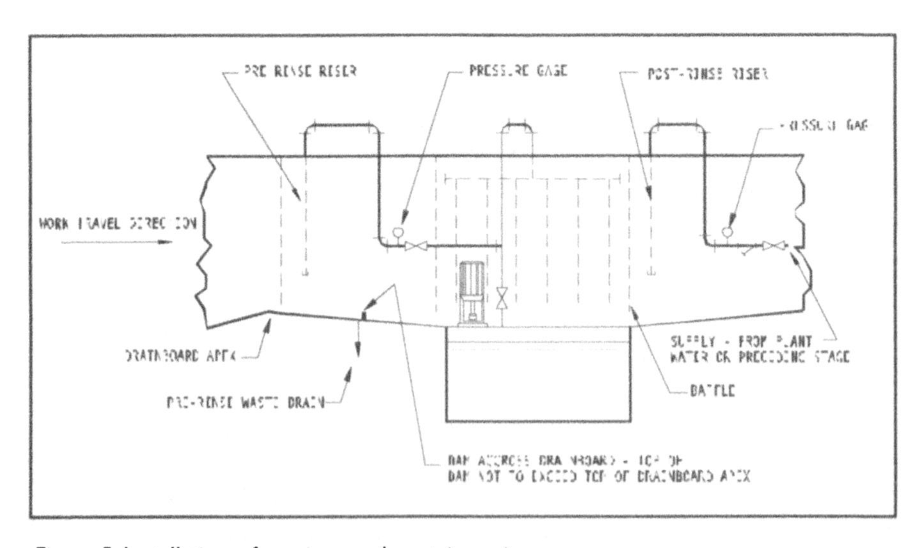

Figure 5. Installation of prerinse and postrinse risers.

ble logic controllers (PLCs), to today's newest personal-computer-(PC-) based "smart control" systems, these advances have provided products in the market with a wide range of control options and quality-monitoring possibilities.

Almost all washer pretreatment processes and their key operating parameters can be monitored and/or controlled with the wide array of field input devices: conductivity, pH, specific ion, and redox potential are all common chemical properties of cleaners and treatments that can be used to control product concentration in the baths. Other devices can be used to monitor or even control the physical parameters of each stage, such as temperature, spray pressure, and even the amount of chemicals fed to the process tanks. The use of proximity switches can allow for work pieces to be counted entering the washer and chemical feed them based on theoretical consumption rates. In short, a control system can be designed to be as simple or as sophisticated as the plant and the chemical supplier feel is required to provide consistent quality and realize the best economical use of washer chemicals.

Once not a concern of most plants, the type of data processor used to build these specialty items is fast becoming the key specification of the control system basis. Customers with other plant operations utilizing one "family" of PLCs will want to keep in-plant spare parts requirements at a minimum, in-plant programming/trouble-shooting scope narrow, and may even want to integrate a pretreatment line's process controller with the plant's communicative network of devices. Recently, plants have been moving towards PC-based systems, because they are increasingly user-friendly, operating in familiar platforms, such as Microsoft Windows, and can be integrated into the plant's Ethernet or Profi-bus communication networks. PC-based systems also tend to be less expensive, but they need to operate in cleaner environments than the heavy-duty PLC-based systems.

Whether PLC- or PC-based systems are ultimately used, the success to any control system lies in the algorithms used to manage all chemical processes. These algorithms, or lines of control logic, must take into account changes in single or multiple analog device signals and determine when appropriate actions should be taken. This is true for chemical concentration and rinse contamination maintenance. Without proper algorithms written for each different line configuration, a control system may not have the capability of delivering optimum control, operating costs, or waste minimization.

A very popular "add-on" to automated chemical feed equipment is data collection or acquisition systems that are capable of storing all measurements of the control system, making that data available for live trending of operating parameters, and archiving that data for future retrieval to investigate problems or longer-term trends. Including these options to a control system allows plants to measure all variables of washer operations and develop statistical analyses for the purpose of meeting their ISO 9000 or QS 9000 requirements.

With such a wide range of possibilities of control options to choose, each system is essentially designed and built for each plant, after thorough discussions. In considering our standard zinc phosphating washer configuration, control systems can cost anywhere between $10,000 to more than $100,000 dollars. Existing lines later adapting control systems have reported washer chemical savings up to 20% when compared with manual feed methods. In general, a design engineer will generally look at the size of the line, its planned production rate, and the anticipated chemical consumption and try to design a control system with a 2–3 year payback. This return on investment does not include opportunities to improve consistency and lower paint rejects. The inclusion or addition of a chemical control system is usually a very effective "upgrading" method in meeting cost and quality improvement objectives.

These are just a few of the more popular ways to approach the process optimization phase in analyzing a washer operation. However, it is in this phase where the creativity of each design engineer will be most apparent, as is evidenced by the wide array of added features to the various washers in use today. However, our work is not quite done, because the waste that will be generated from the washer needs to be addressed and managed.

The "Waste Management" Process

Wastes generated by a pretreatment washer are usually classified in two groups: organic and inorganic wastes. Organic wastes are those oils removed in the cleaning stages, found in smaller amounts in those water rinses after cleaning, and generally seen as undesirable to be carried any further than those rinse stages. Inorganic wastes are found throughout the washer, from cleaning residues metallic and nonmetallic in nature, metallic salts formed as byproducts of the chemical treatment and the subsequent rinse stages, and byproducts of the final sealing chemical rinse, although non-chrome final sealing rinses also may have organic components.

Use of Coalescing and Thermal Oil Separators

The removal of organic contaminants from the work piece, primarily protective oils, lubricants, or machine shop oils, is done in alkaline cleaning solutions designed to remove these soils. If those organic contaminant levels are found to exceed 10,000 ppm (1 vol%) in the degreasing baths, removal of those oils prior to discharge to the general plant waste collection system can be a cost-effective waste management process. The separation and isolation of organic components from entering into the general waste collection stream is a key component of reducing the overall washer discharge volume that will require some form of waste treatment. These oils and greases removed during cleaning are the key contributors to BOD and COD levels in treatment systems.

Coalescing oil separators can be installed as a closed-loop circulation system with the degreasing bath. Oil is absorbed on the media inside the separator tank at the same temperature, or slightly below, normal degreasing bath operation. These separators are quite efficient in their ability to remove oil, and they can be designed for ease in maintenance. In operating at the normal bath temperature of the degreasing solution, this approach to organic contamination removal does not damage the returning cleaner solution in the surfactant's ability to remove additional oils from the work piece. However, depending on the chemistry of the surfactant in the cleaning product and the oils on the work piece, varying degrees of emulsification will take place, with the separator removing oil and surfactant together.

Thermal oil separators work on the same principle as coalescing separators, i.e., using heat and a longer retention time in the off-line, closed loop tank, operating at about 10–40°F above the normal alkaline cleaning solution temperature. This approach is useful only when the surfactant chemistry of the cleaning product is formulated to demulsify under slightly elevated temperatures and when solution kinetics is minimized. In addition, the surfactants used in the cleaning process should not have the capacity to degrade with the high heat applied.

Although design engineers are well versed in the details of these systems, it is the pretreatment chemical supplier that normally recommends the best approach. However, with significant quantities of oils being removed in the degreasing process, the use of either separator system can be cost effective (paybacks of 1–3 years in chemical savings and waste treatment costs) and quality enhancing for those cleaning process.

Use of Ultrafiltration and Microfiltration Systems in Alkaline Cleaning

Ultrafiltration has been used in conjunction with alkaline cleaning solutions for the removal of oils and other contaminants, such as coalescing and thermal oil separators; that is, it is a closed-loop recycling system to reduce chemical use, reduce the amount of spent cleaner required to be auto-drained and/or dumped, and improve overall cleaning performance. However, with a filtering capability down to 10 Å possible in ultrafiltration, removal of key synthetic detergents and surfactants also is possible, which can greatly shorten cleaner bath life.

Recent advances in membrane technology of ultrafilters, as well as investigative work into what components of alkaline cleaners work best with these newer membranes, have resulted in compatible closed-loop systems for minimizing the need to discard spent cleaning solution. If alkaline builders and the type of surfactants (ionic or nonionic) being used in the cleaning process are properly chosen, it may be possible to reduce alkaline cleaning product usage and spent cleaner discharge requirements through the use of these filter systems, while maintaining consistent cleaning performance.

Microfiltration generally filters within a porosity of 0.05–1.0 μm, allowing specially selected detergent builders and surfactants to pass through the membrane and, therefore, be available for reuse in degreasing operations. Microfiltration systems can be designed to process cleaning solutions through multiple passes of filters, coalescers, and, finally, the microfilter, extending bath life 3–10 times that of its unfiltered counterpart. With the consideration of reduced waste, reduced disposal volumes, and washer chemical savings, microfiltration is a practical method of recycling alkaline cleaner solutions, with a reasonable payback period.

Use of Ion Exchange and Nanofiltration Systems with Phosphating Processes

Recently, selective ion-exchange systems have been developed, targeting the removal of heavy metals from water rinses following phosphating processes, where metallic ions, such as zinc, nickel, manganese, and copper, are used. Properly designed, these newer techniques allow for the removal of these metal ions from the water rinse, which later can be recovered and recycled back to the pretreatment chemical supplier for subsequent reuse in new phosphate product concentrate. With new and more stringent environmental laws and effluent limitations being introduced in the next 2–3 years, methods of recycling, using these ion-exchange techniques, may become cost effective when compared with the capital requirements of installing or upgrading waste treatment facilities in plants to remove metals and meet discharge limits. However, at an initial investment cost of anywhere between $100,000 and $1,000,000, only the enforcement of these new regulations may make installing ion-exchange systems on pretreatment lines justifiable.

Nanofiltration systems, large and small, have been developed to work with heavy-metal phosphating solutions in a similar fashion as the above-described ion-exchange systems. Although not capable of reducing nickel and zinc levels down to permissible discharge limits, nanofiltration is a membrane process and not an ionic extraction process through a resin, and it does not require a chemical regeneration to restore filtration capacity, as ion-exchange systems require. A nanofiltration process can be established as a closed-loop collection of rinse water following the zinc phosphating stage, returning heavy metals to the phosphate tank and providing recycled water for all rinsing stages. A significant savings in phosphate chemical use as well as waste water generation can be realized using nanofiltration processes, at one-half or less of initial capital and on-going operating costs, when compared with ion-exchange systems.

Use of Film-Forming Final Rinses

For decades, the standard for final chemical sealing rinses has been chromium based: some have been purely hexavalent based, to target improving coating adhesion, some trivalent chromium based, to provide additional corrosion resistance to the end product, and some a blend of both oxidation states of chromium. Although very effective and easy to maintain on-line, the use of this heavy, toxic metal required special steps in the waste treatment process, creating heavy-metal "filter cakes" that required landfill disposal.

For nearly 15 years, non-chrome-containing alternatives, which have been primarily formulated with proprietary organic compounds, have been available in the market. Some of these non-chrome final sealing rinse processes are reactive in nature and can be subsequently rinsed, while others are nonreactive and only complete the "sealing" of the phosphate coating during the drying process in a low-temperature oven. Although not containing heavy metals, these final rinses do place a further burden on the waste treatment system with the organics they contain, making it more difficult to stay in compliance with BOD and COD effluent discharge limits.

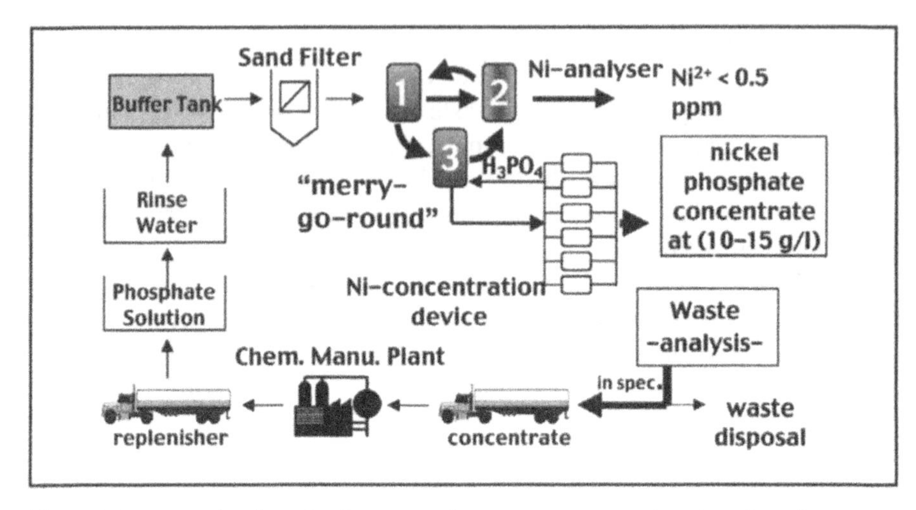

Figure 6. An example of a selective ion-exchange system on a zinc phosphate line.

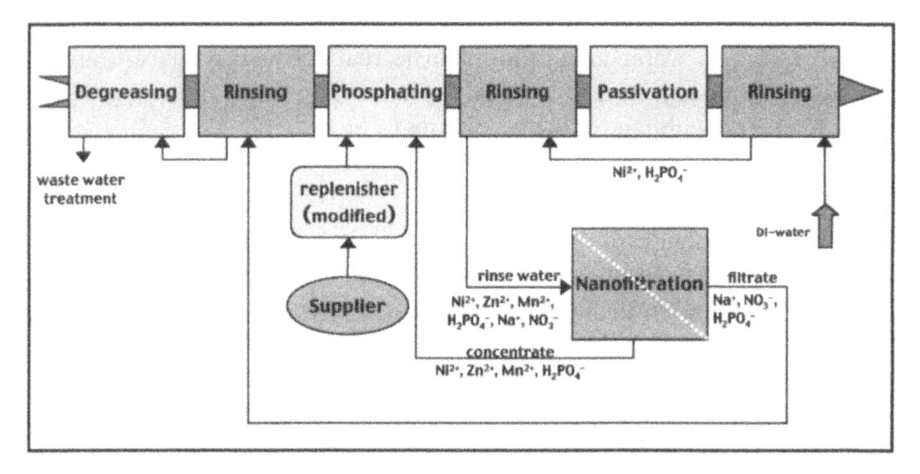

Figure 7. An example of nanofiltration on a zinc phosphate line.

Very recently, newer organic final rinses have been developed that require no rinsing, nor do they require any discarding of bath to a waste collection system to be maintained within contamination limits. These so-called film-forming rinses are actually small organic particles finely dispersed in a misting or "fogging" environment in which the phosphated work piece passes through so that the rinse attaches itself to the surface.

Coating adhesion and corrosion resistance testing results have shown their performances to be equal to the older chrome-containing final rinses, and there is no waste to have to deal with as a byproduct or drag-out.

"Zero-Discharge" Washer Systems

Recently, the market place has been excited by a concept that has been described as zero-discharge washer systems, the cure-all to any waste treatment needs from a pretreatment operation. To clarify the intent of this descriptor, what is actually being described is a pretreatment washer that has been designed to nearly or fully eliminate the need for a continuous discard of chemical baths or rinses to a waste collection system. As long as contaminants are being removed in a cleaning process and a heavy-metal conversion coating process is reacting and, hence, forming byproducts, there always will be generated wastes to address.

It is possible though to eliminate the need for a continuous discharge of wastes and create a means of "batch" elimination of wastes through the use of many of the techniques described here. A list of key elements to designing and successfully operating a zero-discharge washer system may include the following:

- An alkaline cleaning process utilizing closed-loop ultrafiltration to remove oils, with the properly chosen cleaner formulation;
- A system of level controls and counterflows to minimize fresh-water requirements;
- Changing treatment processes from those containing heavy metals to nonmetallic or to an iron phosphate;
- Changing or adding an organic filming final rinse in conjunction with any treatment process change from heavy metals;
- A collection system for a single, low-flow-rate volume of discharge from the system, for subsequent filtering to reuse or to eliminate volume through evaporation; and
- A control system with the proper algorithms in its control logic to manage all processes.

But, here at the end with our cure-all, we have to go back to the beginning and closely watch the performance of our end product, in terms of appearance, corrosion resistance, and overall durability, and evaluate any performance changes with each and every change to our pretreatment process.

Can an existing pretreatment line generating oily and heavy-metal wastes go to a zero-discharge concept without sacrificing quality? Advances in washer cleaning and pretreatment operations, design, and chemical processes described here provide positive feedback to that question with new opportunities.

Conclusion

Meeting the high-performance demands of all coating systems requires an efficient operation of the washer pretreatment process; most of the performance efficiencies will be set based on the intrinsic design of the washer itself. Many waste minimization techniques are used today in the design of a high-efficiency washer; some have been used for many years, whereas others are more recent developments. Although it is always advantageous to initially design new washers with these enhancements, existing washers have been modified to incorporate these cost, quality, and waste reduction improvements. All wastes from a washer and pretreatment operation cannot be eliminated, but they can be significantly reduced or even eliminated from continuously being discharged

Preparation of Cast Iron for Porcelain Enameling

Mike Sexton and Ozzie Storie
Pangborn Corporation

Abstract

The preparation of the substrate for enameling is critical for the quality of the finished article. With sheet steel and other rolled metals, cleaning is normally accomplished by spray washers and cleaning systems. However, with cast parts, and particularly with cast iron, blasting is the preferred method of surface treatment. The process of blasting requires well-designed equipment, an effective abrasive, and a transport system for the substrates for processing through the blasting machinery.

A variety of machines have been designed for abrasive cleaning depending on the type of substrates to be processed. Small parts can be cleaned in batches in units that may be described as barrel units (Figure 1). Large parts can be cleaned in automated systems that may be continuous belts or conveyers traversing through blasting zones (Figure 2).

Figure 1.

Figure 2.

Management of the abrasive is another important aspect of blasting operations. The size, distribution, hardness, and cleanliness of the abrasive have a significant effect on the efficiency of the blasting process. The roughness of the surface can be measured as shown in Figure 3 with surface finish data. A variety of mixes of abrasive are used to tailor the surface roughness (Figure 4).

Approximate Surface Finish Data

(Hot Rolled Steel)

	Surface Profile Height Vertically by Microscope		Arithmetical Average Roughness	Ra (Microns)
Operating Mix	Maximum Mils	Average Mils	Microinches**	Micrometer
S-330	3.3	2.2	300-350	7.62 – 8.89
S-230	3.0	2.1	240-280	6.09 – 7.11
S-170	2.1	1.6	180-200	4.57 – 5.08
*S-110	2.1	1.5	180-200	4.57 – 5.08
*S-70	1.6	1.1	140-180	3.56 – 4.57
GL-18	5.0	3.3	350-500	8.89 – 12.70
GL-25	4.1	2.7	300-340	7.62 – 8.63
GL-40	3.9	2.5	220-270	5.59 – 6.86
GL-50	2.0	1.5	190-240	4.82 – 6/09
GL-80	1.3	0.9	110-130	2.79 – 3.30
G-40CI	3.3	2.2	220-260	5.56 – 6.60

*S-70 and S-110 were full sized operating mixes.

**Arithmetical Average-System of roughness measurement similar to RMS but readings will average approximately 10% less than RMS readings.

Figure 3.

Figure 4.

Figure 5.

After blasting, the abrasive should be cleaned. This can be automatically accomplished with an abrasive separator that uses an air control to separate materials of differing densities and sizes in automatic blasting machines.

During the blasting operation, control of the blasting pattern is exercised by adjustment of the machine settings on a target plate. This is used to optimize the blasting zone or "hot spot" of the abrasive impact pattern (Figure 6 & 7).

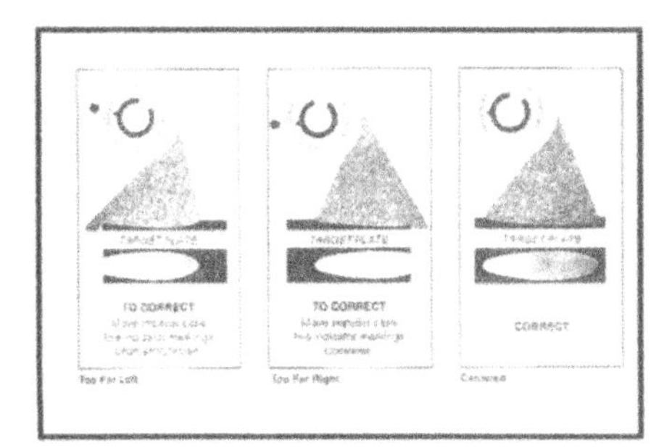

Figure 6.

Figure 7.

Well-regulated machinery and well-chosen abrasive combine to provide an excellent surface finish on cast iron that results in high-quality enamel finishes.

Frit Making Metal Oxides and Regulatory Compliance

Jack Waggener
URS CORPORATION

Abstract

Porcelain enamel glass, known as frit, fried glass in the French language, has been manufactured industrially for more than 100 years. Many of the glass formulations have a long history of successful application and safe consumer use. With increasing knowledge of the physiological and ecological effects of various materials in our daily lives, new regulations are being promulgated at an ever-increasing frequency. The Porcelain Enamel Institute has been active and effective in assisting regulators in developing common-sense approaches to control of various metal compounds used in enamel formulations. Of critical concern is the short-term and long-term exposure of the public and those using these materials through wastewater, air, and municipal waste treatment facilities that handle effluents. Many of the regulated metal compounds, oxides of these metals in many cases—such as nickel oxide, cobalt oxide, copper oxide, manganese dioxide, and molybdenum oxide—and, in more limited cases, chromium compounds, are used in porcelain enamel ground coat formulations. Chromium compounds are of particular concern because of the toxicity of Cr(IV) or hexavalent chrome in mists, dusts, and fumes. This paper illustrate the limits of exposure or control of these materials.

The PEI, the enameling industry, and its consultants have been successful in reducing or eliminating the use of lead, cadmium, selenium, and barium in a variety of enamel systems. When still in use, frits and enameling compounds with these materials are carefully controlled. Ongoing awareness of regulatory issues is important to our industry to ensure that clear understanding of the use and amounts of materials used by our industry is understood by those writing regulations. Enamels continue to be safe and highly effective coating systems for a variety of metal products that produce superior performance for consumers.

Nickel (Ni) (Ground Coats & Oxides)

Wastewater

- Porcelain Enamel ELG
  - 1.41 mg/l Daily Max
  - 1.00 mg/l Monthly Average
- Metal Finishing ELG
  - 3.98 mg/l Daily Max
  - 2.38 mg/l Monthly Average

  POTW (Sludge) – 420 mg/kg

Hazardous Waste (TCLP) – Not Tested For In TCLP Test

Drinking water – No MCL Is Currently Established

Nickel (Ni)

Air Standards

Hazardous Air Pollutant (HAP)

Employee Exposure

OSHA - 1.0 mg/m^3 PEL (8-hour TWA)

NIOSH - 0.015 mg/m^3 REL (10-hour TWA)

ACGIH – 0.2 mg/m^3 TLV (8-hour TWA)

Cobalt (Co) (Ground Coat)

Wastewater

- Porcelain Enamel ELG
  - No Daily Max Limit
  - No Monthly Average Limit
- Metal Finishing ELG
  - No Daily Max Limit
  - No Monthly Average Limit

Hazardous Waste (TCLP) – Not Tested For In TCLP Test

POTW (Sludge) – No Limit

Drinking water – No MCL Is Currently Established

Cobalt (Co)

Air Standards (HAP)

Employee Exposure

OSHA - 0.10 mg/m^3 PEL (8-hour TWA)

NIOSH – 0.05 mg/m^3 REL (10-hour TWA)

ACGIH – 0.02 mg/m^3 TLV (8-hour TWA)

Copper (Cu) (Frit & Oxides)

Wastewater

- Porcelain Enamel ELG (PSES)
  - No Monthly Average Limit
  - No Daily Max Limit
- Metal Finishing (PSES)
  - 3.308 mg/l Daily Max
  - 2.07 mg/l Monthly Average

POTW (Sludge) – 4,300 mg/kg

Hazardous Waste (TCLP) – Not Tested For In TCLP Test

Drinking water – 1.3 mg/l MCL

Copper (Cu)

Air Standards (HAP)

OSHA - 1.0 mg/m^3 PEL (8-hour TWA)

NIOSH – 1.0 mg/m^3 REL (10-hour TWA)

ACGIH - 1.0 mg/m^3 TLV (8-hour TWA)

Manganese (Mn) (Frit)

Wastewater

- Porcelain Enamel ELG (PSES)
  - No Monthly Average Limit
  - No Daily Max Limit
- Metal Finishing (PSES)
  - No Monthly Average Limit
  - No Daily Max Limit

Hazardous Waste (TCLP) – Not Tested For In TCLP Test

Drinking water – No Limit

POTW (Sludge) – No Limit

Manganese (Mn)

Air Standards (HAP)

OSHA – 5 mg/m^3 PEL (8-hour TWA)

NIOSH –1.0 mg/m^3 REL (10-hour TWA)

Molybdenum (Mo)

Wastewater

- Porcelain Enamel ELG
  - No Daily Max Limit
  - No Monthly Average Limit
- Metal Finishing ELG
  - No Daily Max Limit
  - No Monthly Average Limit

POTW (Sludge) – 75 mg/kg

Hazardous Waste (TCLP) – Not Tested For In TCLP Test

Drinking water – No MCL Is Currently Established

Molybdenum (Mo)

Air Standards

Employee Exposure

OSHA - 15 mg/m^3 PEL (8-hour TWA)

NIOSH – 10 mg/m^3 REL (10-hour TWA)

ACGIH – 10 mg/m^3 TLV (8-hour TWA)

Past Problems Reduced or Eliminated by PE Industry

	Haz. Waste TCLP Limit
Lead – PE on Aluminum	**5.0 mg/l**
Cadmium – Reds	**1.0 mg/l**
Selenium – Reds, Yellows	**1.0 mg/l**
Barium – Ground Coats	**100 mg/l**

Chromium (Cr)

Wastewater

- Porcelain Enamel ELG
  - 0.42 mg/l Daily Max
  - 0.17 mg/l Monthly Average
- Metal Finishing ELG
  - 2.77 mg/l Daily Max
  - 1.71 mg/l Monthly Average

POTW (Sludge) – No Limit

Air Standard (HAP)

Hazardous Waste (TCLP) – 5.0 mg/l

Drinking water – 0.1 mg/l MCL

OSHA – Cr (VI)

EXPOSURE TO

HEXAVALENT CHROMIUM

IN

MISTS, DUSTS, AND FUMES

OSHA – Cr (VI)

Some Industry Producers & Users Impacted

Potential Exposed Workers: Over 1 million

- Chromate
- Chromate Pigments
- Portland Cement
- Refractory Brick
- Colored Glass
- Chrome Catalyst
- Porcelain Enamel
- SS Welding
- Electroplating
- Printing
- Chemical Distributors
- Precast Concrete
- Construction & More

OSHA

Evaluating Permissible Exposure Limits (PEL)
to Hexavalent Chromium & Silica

Process:

- OSHA perform studies
- SBREFA Panel Process
  - "Small Business Regulatory Enforcement Fairness Act"
  - Good Process
  - Small Business Critique Pre-Proposal
  - Overseen by: OSHA, OMB, SBA
- Proposed Rule
  - Comments Received by OSHA
- Final Rule

OSHA – Cr (VI)

Under Court Order OSHA is Moving to Greatly Reduce Cr(VI) Permissible Exposure Limit (PEL)

SBREFA	**Winter 2004**
Proposal	**October 2004**
Final	**January 2006**

OSHA – Cr (VI)

Existing PELs

Chromates	100 ug/m^3
Cr(VI)	50 ug/m^3

Cr(VI) PELs being considered

0.25 ug/m3 to 10 ug/m^3

Action Level: 1/2 of PEL

reduction factor: 200 to 5

OSHA – Cr (VI)

SBREFA Panel (February – April, 2004)

Under Estimated Cost Impact

Example

Electroplater
(revenue: $1.4 million/yr)

	OSHA	INDUSTRY
Cost per year	$10K	$160K+
Capital	$1K	$226K

50%+ Plant Closures
Better Science – Health Studies
Reduce to 25 ug/m^3 and .....

Porcelain Enamel Institute Update from Washington on EPA Regulations

Jack Waggener
URS Corporation

Abstract

The United States is undergoing substantial restructuring of its industrial base. Manufacturing jobs are being "lost" to overseas low-labor-cost countries, most notably China. At the same time, imports are exceeding exports by significant margins on a continuing basis. Added to the burden of unregulated international competition are the regulatory environment in the United States and an unequal international regulatory environment. Lowering production costs is forcing manufacturers to seek the low-cost production offered by developing countries and economies.

Regulation of our environment by the Environmental Protection Agency is producing benefits but, at the same time, burdens when regulations are ineffectively or inappropriately applied. Several examples of unneeded and costly regulations "that almost happened" are shown in the accompanying slides. Enforcement of EPA rules has, in some instances, had the appearance of being onerous. This is requiring vigilance by all parties to meet existing standards and manage our materials and processes effectively and efficiently. Care must be exercised to avoid costly mistakes in managing waste streams and materials.

Global Community

Making Sure Regulations Are Fair & We Keep The U.S. Competitive

2.7 million U.S. manufacturing jobs lost since 2000

Increases in Operating Cost

Record Levels of Jobs & Production Pushed Overseas

Part of Solution:

Make sure new Environmental, Safety, and Health Regulations are based on "good" science and are "truly" needed

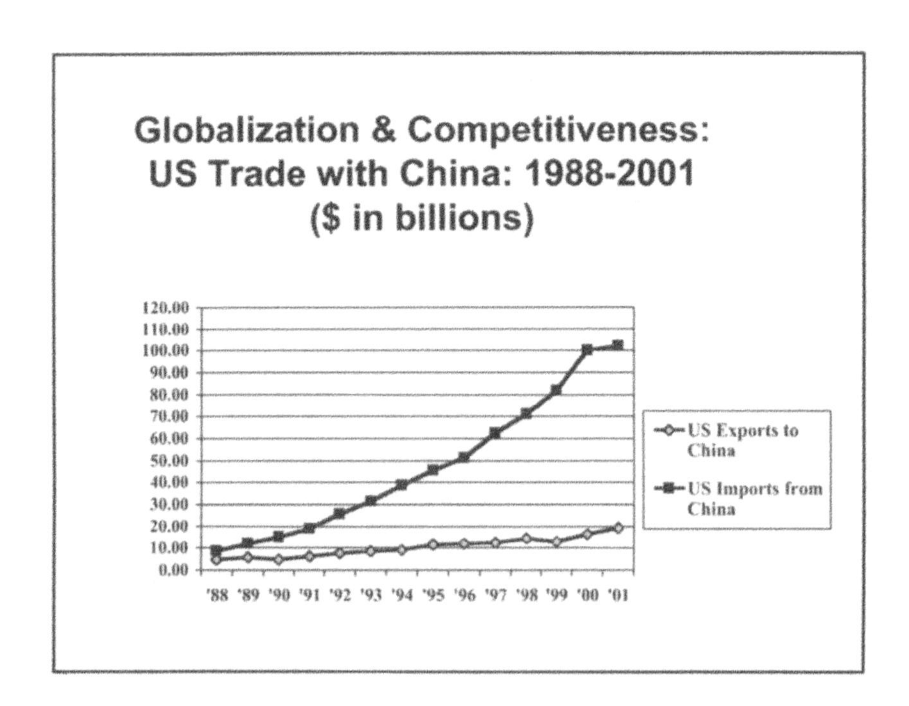

An Example of an Un-Needed and Costly Regulation That Almost Happened

EPA's Metal Products and Machinery (MP&M) Effluent Limitation Guideline

Final Rule: May, 2003
"No Rule"
Avoided Costs: $5 Billion/year

Why: EPA used bad data and science which was shown to overestimate the benefits by a factor of 50 and underestimate the cost by a factor of 10.

This took 10 years.

Some Impacted Industries (100,000 plants)
- Appliances (Porcelain Enamel)
- Plumbing Wares
- Industries producing metal products that are coated, cleaned, fabricated, etc.

304(m) Possible Development of New Porcelain Enamel Effluent Limitation Guidelines (ELG)

304(m) "EPA must periodically re-evaluate existing ELG's

Notice in F.R. (Dec. 31, 2003)

PEI Comments (March 18, 2004)

Toxic Pound Equivalent (PE) Discharged by Porcelain Enamel Users

EPA: "54,077 PE/yr"

PEI Comments: "<875 PE/yr"

Mistakes in EPA Data & Calculations

304(m) (Continued)

Example of EPA Mistakes

Facility "A"

EPA: "52,000 PE/yr" Discharged of Cyanide (CN)

Incorrect flow by 1,350 times

PEI Comment: "38 PE/yr"

Plus facility "A" does not use CN

Facility "B"

EPA: "1,335 PE/yr"

PEI: "Incorrect Flow & Concentration data 550 PE/yr.

Facility has no Porcelain Enamel!!

Stormwater

Construction & Development (C&D-ELG)

Stormwater: Building Expansion of 1 Acre +

F.R.: April 26, 2004

"Withdrew Proposed Rule"

Low Benefits/unnecessary costs

Let Stormwater Phase II work

Let TMDL's Work

X Limits for TSS (10 mg/l.....)

X Post Construction Requirements

X Limits/Sampling

X Restrict to Predevelopment Runoff

Large Cost Avoidance

Watch Out for NRDC Court Appeal & Guidance Doc.

WHAT TO EXPECT

- Redefinition of Solid Waste/Haz. Waste
  Recycling Between Like Industries
  Proposal: Exemption from Haz. Waste
- SPCC – Spill Prevention, Control, and Countermeasure
  New Changes in 2004
- Hazardous Waste: Shop Towels and Wipes/Solvents
  Proposed Exclusion from Haz. Waste
- Air: April 15, 2004: Ambient Air Quality
  470 Countries "Out of Compliance"
  "Nonattainment Areas"
  Expect Air Permit Modifications

Examples of EPA Enforcement

Stored Haz. Waste more than 90 days
$69,000 (PA)
Glass Company demolished 2 furnace Air Stacks without notification
$98,000 (CA)
New Development: Failed to follow stormwater rules
$50,000/day + StopWork (AZ)
Diluting Acid Wastewaters/Diverting Releases from WTP
Facing 20 yrs. Prison
$1 million fine

The Hairline Defect: A Review of the Literature

Robert L. Hyde
Pemco Corporation

Introduction

Since the advent of porcelain enamel on steel, manufacturers have had to contend with hairline defects. Over the years, a great deal has been written on the subject, including theories of their cause and recommendations to minimize or eliminate them. Several test methods also have been developed to simulate the hairline defect in order to study, understand, and determine ways to prevent or eliminate the problem.

Although much has been written about the hairline defect, that information may not be well-known to some of those in the enameling industry. Therefore, the purpose of this paper is to highlight some of the points made in those papers that are still pertinent to today's manufacturers and reiterate the cause and cures previously reported.

Description

Hairline defects, also known as strainlines, primarily occur in the cover coat enamel of a two-coat, two-fire wet enamel system but also can occur in the electrostatic powder system. The defect usually appears as a series of curved thin black lines (ground coat enamel) that generally run parallel to each other. The enamel surface in the defect area is not cracked or fractured but, rather, smooth and continuous. Although they are primarily observed in a cover coat enamel, hairlines can occur anytime a second enamel coating is applied and fired over a previously fired enamel coating.

Cause

The consensus opinion of the cause of the hairline defect is that it originates in the ground coat fire, where certain conditions produce excessive stresses in the fired enamel coating. Early in the cover coat fire (or rework fire), the stresses in the ground coat enamel increase to the point that the coating cracks, which, in turn, causes the bisque cover coat to crack. The ground coat cracking (most likely crazing) at this stage occurs as a result of tensile stress failure. As firing progresses, the ground coat softens enough to "wick" up into the crack of the bisque cover coat because of surface tension.

There are several conditions that contribute to the formation of hairlines, including rough handling, especially if part of the enamel is brushed off; however, today, the parts are handled only minimally. Part design, heating and cooling rates, and uneven heating and cooling can contribute to stresses that lead to hairlines. Areas where a piece of metal has been welded to the back are particularly susceptible to hairlines. The thermal expansion of the ground coat enamel also can contribute to either the formation or the prevention of hairlines. Coating thickness also can contribute to hairlines if not controlled properly.

Papers on Hairline Defects

Four papers on hairlines have been presented at past PEI Technical Forums:

- 1937: "Design and Factors Affecting Hairlining," by E. C. Greenstreet;
- 1955: "The Relationship of Stress and Strain to Processing of Enamels," by J. H. Lauchner;
- 1968: "Controlled Deflection Strainline Test," by G. Clifton Reed; and
- 1970: "Factors which Influence Hairlining of Enamels," by B. J. Sweo and J. D. Snow.

Also of interest are two papers presented in the Journal of the American Ceramic Society:

- 1939: "Observations on Hairlines," by W. A. Deringer; and
- 1954: "A Study of Hairlining," by J. E. Cox and A. I. Andrews, which includes an extensive list of references on hairline defects.

Greenstreet covers all aspects of the process and their influence on hairlines. Aspects include design, forming, milling, pickling, dipping, black edging, and ground coat and cover firing. (Only those suggestions that are appropriate for today's enameling process are mentioned here.) Today, designing parts to avoid hairlines is fairly well understood, especially if the PEI Bulletin P-306 is followed. Therefore, this paper does not include the design recommendations.

The highlights of Greenstreet's paper are as follows:

- A ground coat milled coarser is better than one milled finer;

- A harder ground coat, whether through frit selection or refractory addition, is better (if the bond remains satisfactory);
- A thinner, uniform thickness ground coat is less likely to hairline than a heavy, uneven coating;
- Dipping will produce a more uniform coating than spraying;
- A longer, cooler fire is better than a short, hot fire;
- Avoid heavy tooling and allow sufficient space between parts;
- Combustion products in the furnace can cause hairlines;
- Cover coats generally follow the same guideline as for the ground coat;
- Keep the cover coat thickness to a minimum so long as the coverage is sufficient;
- Use a cover coat formula that has good bisque strength; and
- Spray cover the coat on the back of those areas that are likely to heat up faster.

Lauchner measured the deflection of several enameled test panels over a temperature range and discussed the characteristics of an annealed versus an unannealed glass and the associated stresses. He theorized that the stresses in a ground coat enamel are the result of the cooling rate of the ground coat fire. Two studies were conducted: one in which a box fire was compared with a continuous fire and one where the cooling rate was varied in the continuous fire. Test panels used in this study were prepared using 14 gauge, 6 × 8 in. plates, with a 2 × 6 in. plate that were spot welded to the back. The firing cycle for each furnace was adjusted in an attempt to produce an equally fired enamel coating. The hairline defects were worse from the box furnace compared with the continuous furnace. (This would be expected, because fast heat-up and cool-down of the ware is unavoidable.) It follows then that the rapid cool-down cycle typical of a box furnace fire aggravated the hairline problem. In the second study, as the cool-down rate was decreased, hairline defects were decreased or eliminated. This study confirmed that hairline defects could be decreased or eliminated by increasing the cool-down time in the ground coat fire, which allowed the stresses to decrease to an acceptable level.

Reed reviewed the current test methods (available in 1968) for measuring the hairline resistance of an enamel coating and presented a new test method. The author indicated that the current test methods at that time were not sensitive enough to help evaluate new frits developed to be more hairline resistant.

Sweo and Snow described the stresses in an enamel that may lead to hairlines if not kept within certain limits. Two forms of stress were discussed: inherent stress, which results from the difference in thermal expansion between the steel substrate and the enamel coating; and external stress, which results from temperature differences within the test sample (or production part). They were able to correlate deflection to stress in ground-coated test panels for two enamel formulas of higher and lower fusibility using a temperature deflection test (similar to that described by Lauchner). Through experimentation, they were able to determine stress values of both enamels over a temperature range. They found the following:

- There is a relationship between the inherent stress and the applied stress in a fired enameled part;
- The applied stress required to cause hairlines decreases as temperature increases; and
- The softer enamel in the study was more prone to hairlines because of its higher thermal expansion, which caused it to be in greater tension during the cover coat fire.

Deringer discusses seven factors that influence the formation of hairlines and offers recommendations for each:

- Design.
- Ground coat formula. A good ground coat is one of the strongest defenses against hairlines. Increasing the refractory (silica or feldspar) content tends to improve hairline resistance.
- Firing conditions. Two schools of thought are presented. (1) A hard-fired ground coat is less likely to cause hairlines than one that is underfired. (2) Fire the ground coat to a deep blue at a relatively low temperature. (In 1939, most, if not all ground coat frits were high cobalt.) The thought here is that a lower, longer fire will distort the wear less. Uniform heating of the wear may be more important than the firing time or temperature.
- Heating and cooling rates. Sources of strain sufficient to cause fracturing are discussed. The sources include fabrication induced, flexing, and unequal rates of expansion or contraction during firing.

How the wear is positioned on the furnace chain may help or hurt hairline resistance. Minimizing uneven heating of the wear is key to minimizing hairlines.

- Ground coat thickness. The thinner the coating, the better and the more uniform the ground coat thickness.
- The substrate. The steel substrate should have excellent sag resistance.
- Cover coat enamel. Cover coat frits in 1939 were mostly antimony or fluoride opacified. Therefore, the recommendations may not apply here.

Cox and Andrews presented an extensive paper that included a test panel similar to that of Lauchner. The panel was developed to induce and, thereby, study the development of the hairline defect. Hairline formation occurs very early in the cover coat fire. Stresses develop because of distortion of the part, and cracking of the ground enamel occurs, which in turn causes the cover coat bisque to crack. Factors that improve hairline resistance include ground coat refractory mill additions, dense ground coat bubble structure, and thin coating. For the cover coat, fluidity and thickness are most important. The authors showed that the addition of silica (325 mesh) did not eliminate ground coat cracking but, rather, caused the cracks to be much shorter and discontinuous, a condition that is much better tolerated by the cover coat, thus minimizing or eliminating hairlines. Rapid cooling of the ground coat is more likely to cause hairlines than is slow cooling, as related to the stresses remaining in the ware after the ground coat fire. A ground coat thickness greater than 5 mils hairlined, whereas thicknesses less than 5 mils did not hairline. The authors also identified the critical areas in the firing cycle that contributed to the formation of hairlines. From this, they presented a firing cycle that should minimize or prevent the formation of hairlines.

Conclusion

The hairline defect, when it occurs, can be costly for the manufacturer through either rework or scrap. Therefore, it may be worth a little time to review the process and consider the following key points from these papers:

- Part design is very important;

- A longer, lower, and uniform ground coat fire will minimize stresses in the ground coat;
- Coarser ground coat and cover coat slip fineness will help minimize hairline defects;
- A harder ground coat is less likely to hairline than a softer one because of the generally lower thermal expansion of harder ground coats, which are less likely to be placed in tension during firing; and
- Keep the ground coat thickness to a minimum and, to a lesser degree, the cover coat.

The Physical and Chemical Characteristics of Porcelain Enamels

William D. Faust
Ferro Corporation, Cleveland, Ohio

Introduction

For more than 100 years, porcelain enamels have been the most durable coating on metal for many domestic and industrial applications. The unique properties of glasses are an integral part of porcelain enamels when applied to metal substrates. Additionally, the effect of compression of the glass–metal composite after firing adds to the physical and thermal properties. Design of appliances and other products today need to take advantage of the properties that can be developed but not to exceed the performance characteristics that tend to detract from the application of porcelain enamels.

The Chemistry of Porcelain Enamels

The make-up of porcelain enamels is normally that of an alkali–borosilicate glass. Compositionally, enamels can be characterized as follows:

- Quartz (SiO_2) 35% to 60%;
- Borax ($Na_2O{\cdot}2B_2O_3{\cdot}10H_2O$) 10% to 15%;
- Feldspar ($K_2O{\cdot}Al_2O_3{\cdot}6SiO_2$) 10% to 15%;
- Minor ingredients 10% to 20%; and
- Metal oxides, fluxes, etc.

Porcelain enamel glasses are generally lower-temperature glasses when compared with other types, such as window glass and cookware glasses. Various properties can be imparted to the glasses through the introduction of various metal oxides to develop self-adhering enamel and self-opacifying enamel, such as appliance white coatings. Metal oxides are essential for the development of fired adhesion of the glass to the metal. The mechanism of adhesion is a chemical–mechanical bond that develops during the firing process.

The firing process is a complex interaction between the glass and the metal substrate. Dissolution of ferrous metal at the glass–metal interface is critical for adhesion development. This reaction produces a haze layer at the interface, which is normally an unattractive color. Additional enamel thickness is used to develop the physical, chemical, and esthetic properties of the glassy coating.

Thermal Properties of Glass

Glass is characterized by viscosity. At room temperature, the viscosity of glass is extremely high and we do not observe any changes in properties, hence, the usefulness of glass coatings and objects. As exposure to increased temperatures occurs, the viscosity of glass changes and the properties change. Figure 1(A) illustrates a generalized viscosity diagram for an alkali–borosilicate glass.

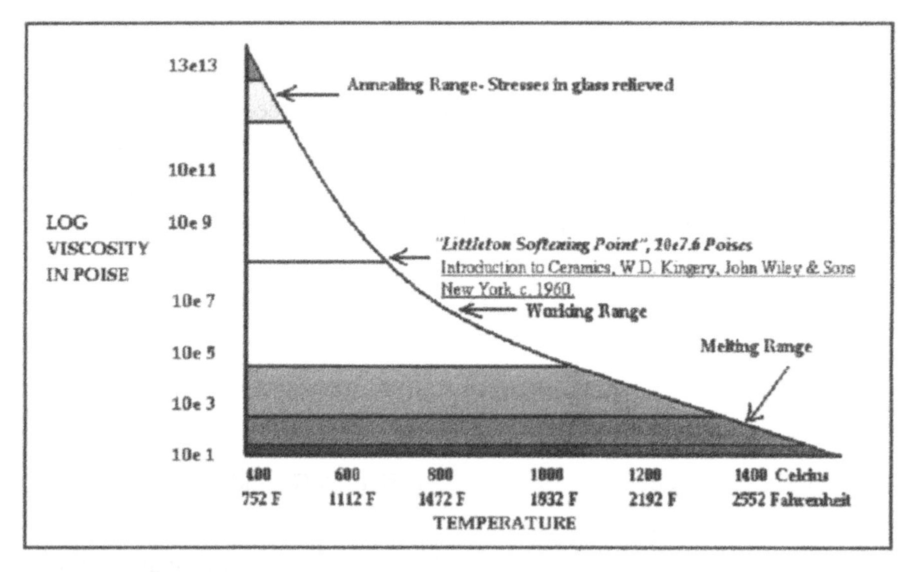

Figure 1(A). Viscosity versus temperature for alkali–borosilicate glasses.

Porcelain enamels exhibit the same characteristics when applied and fired on metal substrates. Additionally, the mismatch in the thermal expansions of the glass and the metal results in compression of the glass at room temperature and at moderately elevated temperatures. Figure 1(B) illustrates the effect of the different thermal expansions between steel and two enamels with different expansion values. Lower-expansion enamels develop higher compression values; however, they also are harder glasses that

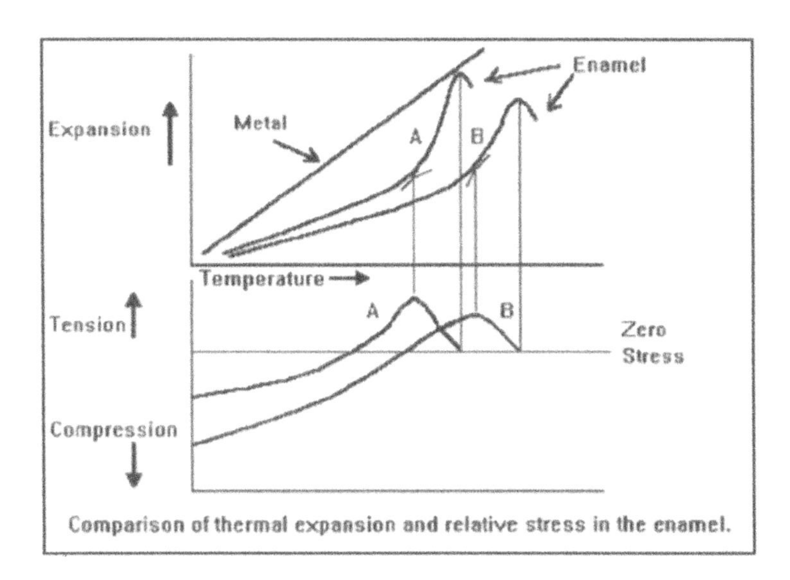

Figure 1(B). Thermal expansion and stress development in enameled steel. (From: *Kirk–Othmer Encyclopedia of Chemical Technology,* 4th ed., Vol. 9. Wiley, New Work, 1999.)

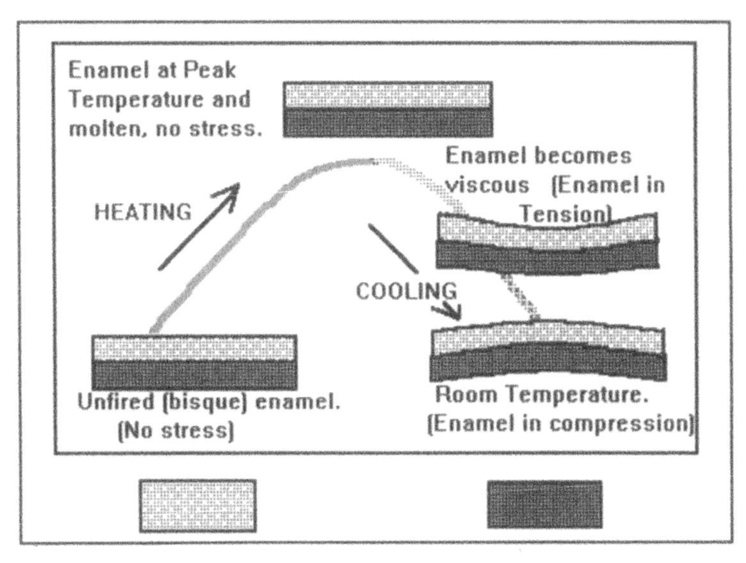

Figure 2. Stress–strain relationship in enamel and steel (general characteristics). (From: William D. Faust, "Practical Considerations Regarding Stress and Strain in Enamel–Steel Composites"; p. 175 in *18th International Enamellers Congress* (Paris, 1998)).

have limitations regarding firing temperatures. Figure 2 illustrates the effect of the thermal expansion mismatches and the resultant compression and tensile forces. The enamels are stress-free before firing and up to their firing temperatures. Upon cooling, enamels first exhibit tensile stresses because of the higher rate of contraction of the glasses than that of the metal and then compression as the cooling continues to room temperature and the rate of contraction of the glasses slows down and is exceeded by that of the metal. Steels generally have near linear thermal expansions in the temperature ranges of normal enameling.

Chemical Properties of Glasses

Glasses have structures that are distinctly different from other materials. Many materials exhibit crystalline structures with regular arrangement of atoms. Crystals are frequently characterized by having specific melting points. This regular arrangement of atoms minimizes the ability of other atoms or ions to enter the structure of these materials. Quartz is an example of a crystalline material that has a strong and regular structure for a ceramic material and exhibits structural stability in various polymorphic forms from about 573°C (1063°F) and higher. Glasses have amorphous structures that

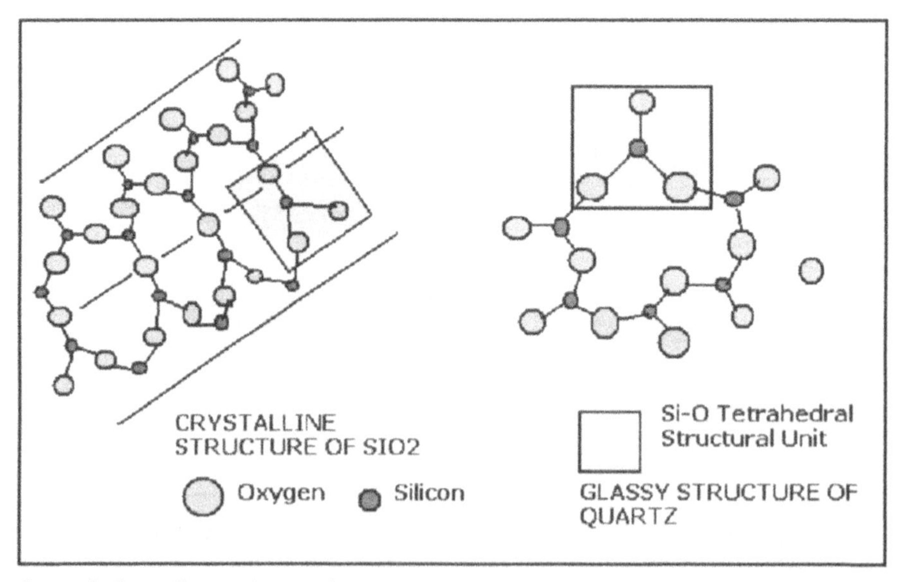

Figure 3. Crystalline and amorphous structures of materials. (From: W. D. Kingery, *Introduction to Ceramics;* Fig. 5.3, p. 144. Wiley, New York, 1960.)

do not exhibit distinct structures at any temperature. X-ray diffractions of glasses and enamels show no distinct patterns of peaks. However, glasses have the ability to absorb many different types of atoms into their structure that, in turn, allow the creation of a multitude of material properties. Figure 3 illustrates the difference between the regular structure of crystalline quartz and that of amorphous or glassy quartz, an example of which is fused quartz or obsidian (volcanic mineral).

The glassy structure of glass and enamels allows for a wide range of material properties as well as chemical durability. Various alkali ions (Na^+, K^+, and Li^+) and alkali-earth ions (Ca^{2+}, Sr^{2+}, and Ba^{2+}) are frequently used to make glasses with a wide range of workability. The alkalis decrease the melting and application temperatures to reasonable levels, and the alkali-earths impart durability without undue refractoriness.

Chemical durability is affected by the interaction of the solute and the surface of the glass. The alkali components are initially selectively removed followed by gradual breakdown of the silica skeleton. The corrosion mechanisms of porcelain enamels were studied by Eppler *et al.*[1–4] Two processes were indicated. One was that "an ion-exchange process may occur between a modifier (principally alkali) ion in the glass and a hydrogen ion in solution:"

$$H^+ \text{ (solution)} + Me^+ \text{ (glass)} \rightarrow H^+ \text{ (glass)} + Me^+ \text{ (solution)}$$

In this case, in an acid environment, the ion exchange occurs without disturbing the glass structure; that is, the silica network remains. Moreover, the silica-glass network may be dissolved through attack by alkali hydroxides that break down the Si–O–Si structure:

$$\text{Si–O–Si} + \text{NaOH} \rightarrow \text{Si–O–Na} + \text{Si–OH}$$

These processes are dependent on the pH of the solutions below a pH of 9. Corrosion of water tank enamels also was studied. Along with pH, the formation of surface complexes on the glass–liquid interface decreases the corrosion rate because of adherent layers. This may occur because the surface layer decreases the activity of the solution or the surface layer interacts with the hydrous silica ions on the glass surface.

In many instances, refractory materials are combined with the enamel formulations in the mill to create a complex coating containing discrete particles of silica (quartz), zirconium silicate (zircon), feldspar, and nepheline syenite or other refractory materials. These materials enhance the corrosion resistance of enamels and impart improved abrasion resistance in many instances. Specially formulated water heater enamels may contain up to 40% mill-added silica to create very durable coatings.[5] Improvement in the acid, alkali, and abrasion resistance is possible with the addition of silica to a standard ground coat system. However, the bond development and the optimum fire or maturing temperature increases as more refractory is added.[6] About half of the mill-added silica can be dissolved into the glass matrix.[6] The remainder is encapsulated. The dissolved silica may alter the glass formula sufficiently to improve the chemical properties. The undissolved particles may help provide abrasion-resistant areas in the glass–silica matrix. Zircon has been shown to improve the alkali resistance as well as providing an increase in compression of the enamel.[7] Again, some dissolution of the zircon is likely in changing the glass chemistry as well as providing some physical resistance to abrasion by the larger undissolved particles. Figure 4 shows an enamel ground coat with silica (quartz) and zircon particles in the glass matrix.

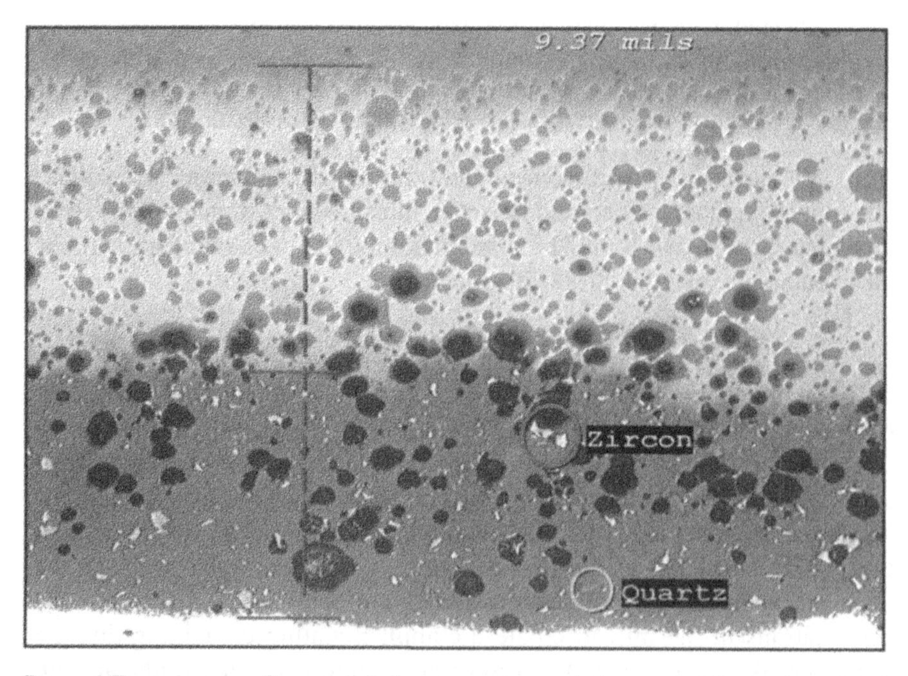

Figure 4. Two-coat–two-fire enamel showing quartz and zircon particles in the ground coat glass matrix of a fired coating. A titanium opacified cover coat is seen above the ground coat.

Table I. Sheet Steel Porcelain Enamels—Typical Physical Properties

Property	Value
Density (g/cm^3)	2.4–2.7 (variable because of composition)
Specific heat (g·cal/°C)	0.20–0.30
Conductivity (g·cal/(cm^2·s)	1.812–3.243 at 100°C; 1.698–2.796 at 0°C
Conductivity (W/(°C·cm)),	3/19/03
second source, no reference	0.05–0.05
Heat of fusion (Btu/lb)	Net endothermic reaction for melting soda–lime glass, 81.3
Fusion temperature (°F(°C))	1400–1500 (760–815) (for porcelain enamel coating on steel)
Young's modulus (psi)	8.0y × 10^6 (typical)
Poisson's ratio	0.20–0.24 (0.22 typical)
Residual stress (psi)	15×10^3–30×10^3
Thermal expansion (in./(in.·°C))	80×10^{-7}–100×10^{-7} (for acid-resisting enamels, closer to
	80×10^{-7}–85×10^{-7})
Moh's Hardness	5–6
Tensile strength (MPa (psi))	34–103 (4900–15000)
Compressive strength (MPa (psi))	1380–2760 (2×10^5–4×10^5)
Modulus of elasticity (GPa (psi))	55–83 (8×10^6–12×10^6)
Dielectric constant	5–6
Strain in enamel that leads to failure (cm/cm)	0.002–0.003

From: A. I. Andrews, *Porcelain Enamels.* The Garrard Press, Champaign, Ill., 1960. Kirk–Othmer Encyclopedia of Chemical Technology, 4th Ed., Vol. 9, *Enamels, Porcelain or Vitreous,* pp. 413–38.

Application Characteristics

Porcelain enamels are used in many appliances because of the mechanical, thermal, and chemical durability they provide. Mechanically, porcelain enamels stiffen the glass–steel structures by virtue of their lower thermal expansion compared with that of the steels. If a metal part is coated on only one side, the part may exhibit a "warp" or bend because of the bimetallic effect of the expansion differences between the two types of material. Coating both sides of a thin metal part equalizes the stresses and produces a more planar product. The built-in compression stress provides added strength in that the compression must be overcome before fracturing of the

Table II. Engineering Properties of Steel

Property	Value
Young's modulus of elasticity, E (ksi)	29,000
Fatigue endurance limit, FEL, reverse bend test, R (cycles)	–1, 10^7
Low- and medium-carbon grades	FEL = 47% of UTS†
HSLA grades	FEL = 44% of UTS
Physical properties at room temperature	
Density (lb/in.^{3})	0.283
Coefficient of thermal expansion (in./(in.·°F) (in./(in.·°C))	6.7×10^{-6} (120×10^{-7})
Physical properties from room temperature to 200°F	
Thermal conductivity (Btu/(h·ft·°F))	35
Specific heat (Btu/(lb·°F))	0.111
Electrical resistivity (μΩ·in.)	5.5

From: AK Steel Monograph, 1995.

†UTS is ultimate tensile strength.

glass coating occurs. Thinner coatings have higher compressive stresses as do glasses with lower thermal expansion coefficients.

Utilization of the physical properties of enamels in the design stage of producing a product assists in optimizing the application, particularly if computer-aided design is used to apply the material properties.

The properties of steel, the most common substrate for porcelain enameling, are critically important to the designer and to the producer of enameled articles.

In addition to the fundamental material properties, porcelain enamels are unique in their application. As shown in Figure 5, the enamel coating is generally in compression if it is not excessively heavy. Very heavy coatings may revert to a state of tension on the surface and be prone to chipping.

If the enamel tends to become too heavy, that is, more 15 mils, edge chipping is often observed. This is seen on the edges of burner caps, range corners, and any other areas that may accumulate more enamel. Recoating of parts often leads to excessive thickness if the second coat is not kept to a minimum.

Other properties of enamels are well documented in the various Porcelain Enamel Institute publications.

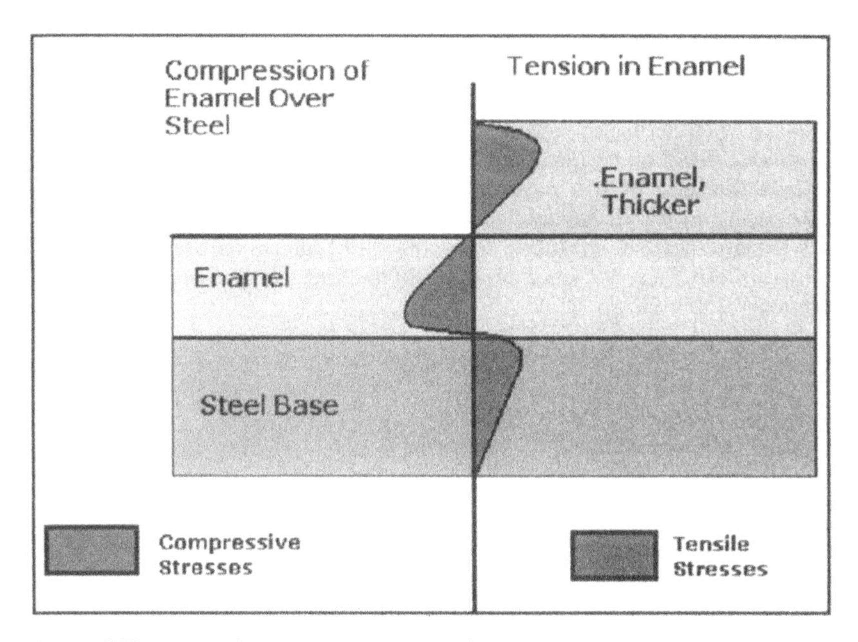

Figure 5. Tension and compression in enamel coatings.

Conclusions

Enameling has specific fundamental properties that can be used to make better and more functional products. The long history of enameling has provided important information for successful development of the enamel–steel composite, whether it uses steel, cast iron, stainless steel, aluminum or some other metal.

References

1. Richard A. Eppler, Robert L. Hyde, and Howard F. Smalley, "Resistance of Porcelain Enamels to Attack by Aqueous Media: I. Tests for Enamel Resistance and Experimental Results Obtained," *Am. Ceram. Soc. Bull.,* **56** [12] 1064–67 (1977).
2. Richard A. Eppler, "Resistance of Porcelain Enamels to Attack by Aqueous Media: II. Equation to Predict Durability," *Am. Ceram. Soc. Bull.,* **56** [12] 1068–70 (1977).
3. Richard A. Eppler, "Resistance of Porcelain Enamels to Attack by Aqueous Media: III. Mechanism of Corrosion of Water Tank Enamels," *Am. Ceram. Soc. Bull.,* **60** [6] 618–22 (1981).

4. Richard A. Eppler, "Resistance of Porcelain Enamels to Attack by Aqueous Media: IV. Effect of Anions at pH 5.5," *Am. Ceram. Soc. Bull.,* **61** [9] 989–95 (1982).
5. Howard J. Smith, "The Effect of Refractory Mill Additives on the Corrosion and Abrasion Resistance of Porcelain Enamels"; pp. 37–48 in *Proceedings of the Porcelain Enamel Institute Forum,* Vol. 15, 1963.
6. Lester M. Dunning, "High Refractory Mill Additions in Ground Coat"; pp. 34–36 in *Proceedings of the Porcelain Enamel Institute Forum,* Vol. 15, 1963.
7. R. B. Kempson and J. R. Friedrichs, "Refractory Mill Additions for Use in Sheet Steel Ground Coats"; pp. 49–54 in *Proceedings of the Porcelain Enamel Institute Technical Forum,* Vol. 15, 1963.

The Effect of Refractory Mill Additions on the Thermal Expansion of Enamel

Boris Yuriditsky
Pemco Corporation, Baltimore, Maryland

Introduction

Crazing, spalling, warp, and other problems are caused by stress that develops between enamel and steel substrate during the cooling cycle of the firing process. This effect is due to the differences in coefficients of thermal expansion (CTEs) of the two materials. Above the glass transition temperature (T_g), in the process of cooling, the stress is relieved first by the viscous flow and, at the lower temperatures, by the plastic deformation of the enamel.

The stress begins to build up when the glass viscosity increases to such a level that the deformation of the enamel changes from plastic to elastic. The temperature of this transition is close to the T_g. T_g can be measured from the glass expansion curve (Figure 1).

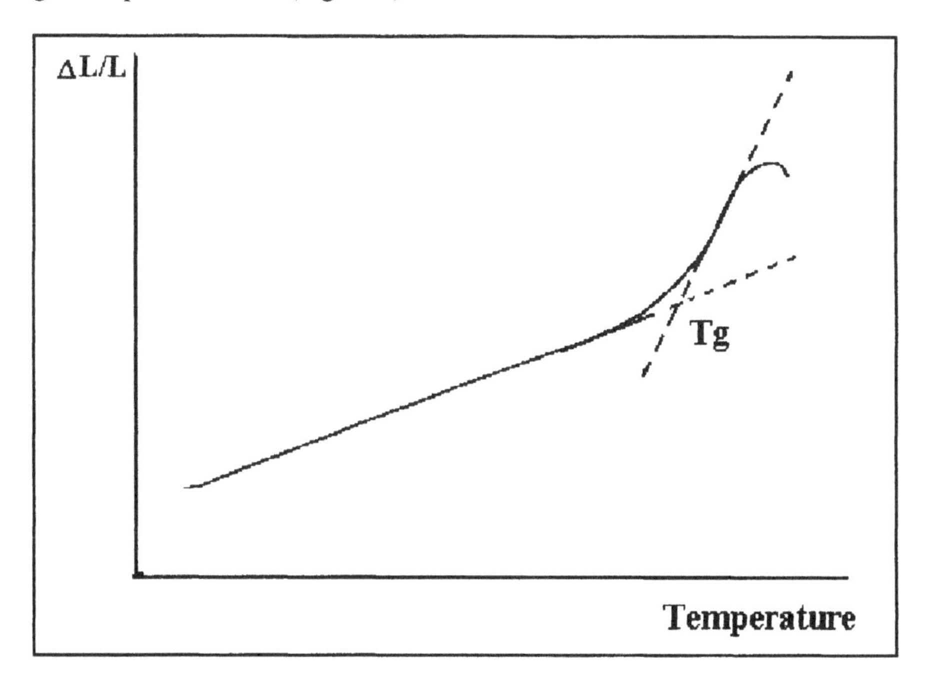

Figure 1. Thermal expansion curve of glass.

The incipient fusion point (IFP), T_g, and CTE are functions of the glass composition and, therefore, are related. If we consider these factors independent and everything else equal, the difference in stress of an enamel at room temperature (20°C) that has a T_g of 500°C (typical) and an enamel with a T_g 25°C lower is (500 – 20)/(475 – 20) ≈ 105%. In other words, a 25°C increase in Tg increases compressive stress as much as it decreases CTE from 315 × 10^{-7}°C^{-1} to 300 × 10^{-7}°C^{-1}. From practical experience, this is a very substantial difference that should be considered (Figure 2).

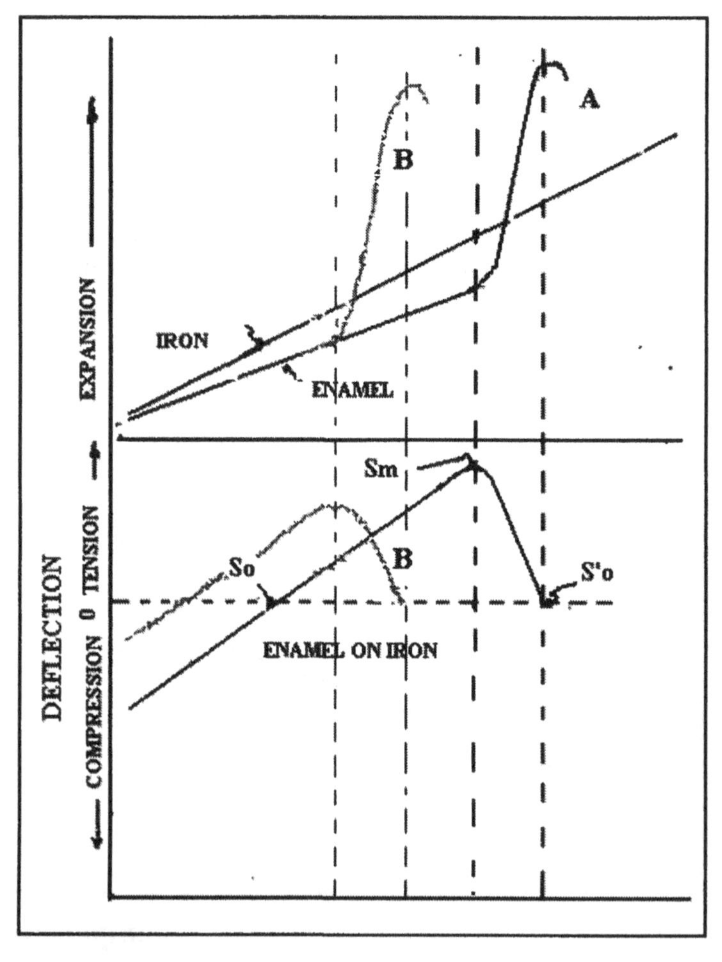

Figure 2. Inherent stress in enamels A and B in relation to thermal expansion of enamel and iron.[6]

Actually, the difference in stress (σ) is even more substantial, because the glass with a lower T_g is usually softer. Softer glasses usually have lower moduli of elasticity (E). Because σ is directly proportional to E, a decrease in E decreases the stress. Also, an increase in T_g decreases maximum tensile stress at σ_m, which also contributes to the increase in compressive stress.

The cause of hairline is tensile stress. In addition to the stress that develops as a result of thermal gradient in a ware, stress develops at the cooling of the system from σ_o (IFP) to σ_m (T_g), where the enamel CTE is higher than the steel substrate CTE. IFP is the point where enamel begins to exhibit plastic deformation. The higher the T_g, the lower is the maximum inherent stress at σ_m. Lower stress at sm corresponds with higher so temperature and the wider range where glass is safely under compressive stress (Figure 2). Therefore, to improve crazing resistance, T_g should be increased. Practically, this is achieved by making a glass harder. Even in this simplified model, the CTE of the glass cannot describe stress development in the enamel. The initial tensile stress above the T_g point is a function of CTE above T_g and IFP. These parameters are not directly related to the low-temperature CTE.

It is impractical to address each stress-related problem by developing a new frit. Stress can be controlled by the addition of refractory materials in the mill formula. Some of these additions dissolve in glass slowly and mostly stay in the enamel as a separate phase. We refer to these additions as refractory to distinguish them from other high-temperature oxides, such as calcium and magnesium oxides that easily dissolve in glass.

Many factors other than CTE, IFP, and T_g influence stress in an enamel–steel system. The properties of steel—E, CTE, and thickness and shape of the substrate—are among them.[7]

The structure and the composition of enamel also are very important. Defects, such as microcracks and bubbles, relieve stress. The iron oxide penetration layer on the enamel–substrate interface creates a transitional layer, which is different from the volume of enamel. The relief of stress through the plastic deformation of enamel is a time-related process; therefore, the speed of cooling also is a factor in stress development. These, along with many other factors, are beyond the scope of this study.

The purpose of this work is to study the effect of mill-added refractory on T_g and CTE of the enamel.

Table I. T_g of Ground Coat Enamel with 20% Refractory, Fired at Different Temperatures

	Tg (°F)				
T (°F)	As is	Feldspar	ZrO_2	TiO_2	$ZrSiO_4$
1400	298	292	269	276	257
1500	298	286	274	287	253
1600	299	289	270	284	249

Procedure

The most common refractory materials used as mill additions are silica (SiO_2), titania (TiO_2), zircon ($ZrSiO_4$), zirconia (ZrO_2), and feldspar. All these materials are present as a separate phase in fired enamel.

The mineral composition of feldspar is 57% albine, 6.1% anorthite, 24% orthoclase, 10.2% silica , and the balance mica.

Density and CTE of this composition were calculated based on the individual data from References [1–3].

In ceramic practice, volume CTE is used to avoid the problem of anisotropic thermal expansion. Consequently, where not stated otherwise, mill additions are expressed in volume percent, as a percentage of the frit.

The mill formula is 100 g frit, variable refractory, 3 g ball clay, 0.3 g bentonite, 0.2 g $NaAlO_2$, 0.2 g K_2CO_3, and 45 g H_2O.

The clayless, bubble-free mill formula is 100 g frit, variable refractory, 0.02 g Peptapon 99, and 45 g H2O.

The composition was milled to a fineness of ~6%/200 mesh.

The material was dried and fused into bars about 4 mm × 4 mm × 50 mm in nickel containers at 1500°F (if not stated otherwise) for 12 min. At the cooling cycle, the bars were annealed with soak temperatures of 482°C for 60 min, 454°C for 30 min, and 399°C for 30 min, with a ramp speed of 2°C/min.

Results and Discussion

In all studied ranges of materials and concentrations, there is little variation in the value of T_g. Everywhere, T_g is ~500°C ± 10°C. There is some trend for TiO_2 to increase in Tg from 495°C to 505°C–510°C. Feldspar tends to decrease in T_g to 485°C–490°C. No connection between T_g and porosity or amount of refractory is detected. Consider that the thermal expansion curve is not perfectly linear; the difference in T_g is within normal variation. If the process of refractory dissolution in the glass would be significant, it would be reasonable to expect a significant change in T_g.

Only the general observation that addition of refractory somehow decreases CTE could be made on compositions with developed bubble structure (clay-containing system). The variability of the measured CTE is so great that no meaningful relation between refractory content and CTE of the enamel could be found. Decrease in clay content does decrease variability, however. Only bubble-free samples yield meaningful results. If one were to regard bubbles as inclusions in enamel, than size and concentration of bubbles should influence the CTE. Consider that bubble structure in the relatively thick specimens is hard to control; high variability of results should be expected.

The study was done on semiopaque and ground coat frits. The former was selected as a frit where various refractory materials are often added and the latter as a generic ground coat frit where stress control is important.

No meaningful results were obtained on semiopaque frit. It was found that refractory particles promote crystallization of TiO2 from the semiopaque glass at an unpredictable rate. Additional TiO2 crystals contribute to thermal expansion, which causes the result to vary.

There is no significant difference in CTE between materials fused at 1400°F, 1500°F, and 1600°F (Table I). Possible explanations could be that no or little dissolution of refractory in glass takes place in the studied temperature range or that the difference in CTE between heterogeneous system refractory–glass and the homogeneous glass of the same composition is small.

After many trials, it was established that variation in CTE of SiO_2-containing enamel was much higher than in other compositions. For all systems with the exception of SiO_2, CTE variation was ±1.5%–2%. For SiO_2, the variation was ±4%–5%. Even the well-known fact that mill-added SiO_2 increases hairline resistance and promotes crazing cannot be correlated with the thermal expansion data. X-ray diffractometry of the slowly annealed specimen and the water-quenched specimen shows that only α-quartz exists in all temperature ranges. There is no SiO_2 phase transition that would make the results vary. Probably, unlike other refractories, SiO_2 reacts or/and dissolves in glass with an unpredictable rate. This process significantly influences CTE.

The CTE–refractory-content relation for other refractories is close to linear (Figure 3).

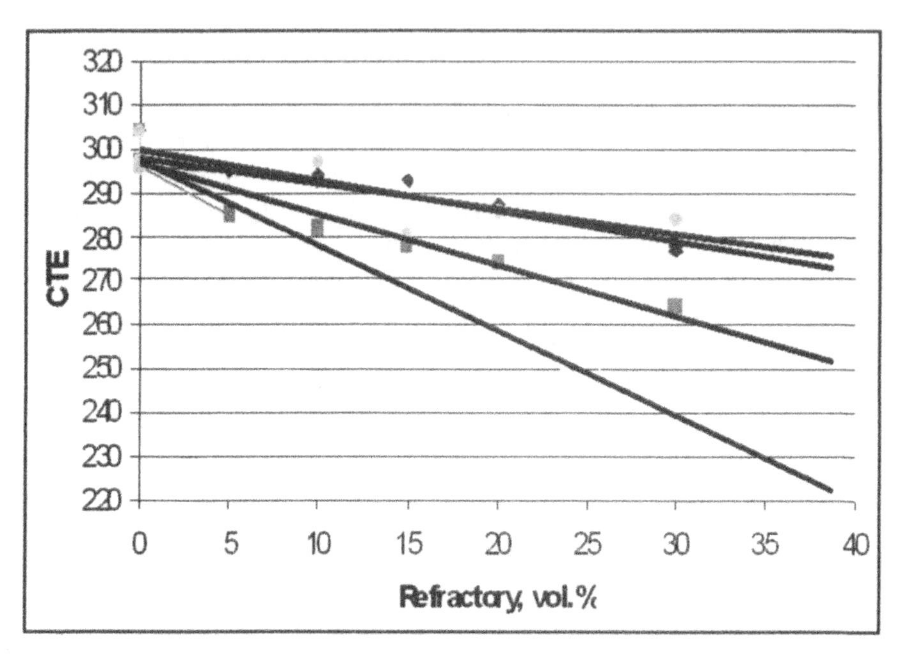

Figure 3. CTE versus refractory content.

The relative effect of refractory on CTE does not correspond with CTE of the individual refractory materials. (Table II), as was expected. The CTE of the phases is just one of the factors in the CTE of the composite. The others factors that pertain are modulus of elasticity and Poisson's ratio. The CTE of composite material is given by[4]

$$\alpha_r = \frac{\sum_1^i \alpha_i K_i V_i}{\sum_1^i K_i V_i}$$

where $K = E/3(1 - 2\mu)$, E the elastic modulus, μ the Poisson ratio, α the volume expansion coefficient, and V the volume part of the phase. Nevertheless, this data can be useful in the estimation of the composite enamel CTE.

Also, the CTE of crystalline materials is a nonlinear function of temperature. For example, for MgO from 50°C to 500°C, CTE changes from ~200 $\times 10^{-7}$°C^{-1} to ~390 $\times 10^{-7}$°C^{-1}.[1]

Table II. Coefficients of Regression $\alpha = \alpha X + b$, where X is Refractory Content (vol%)

	CTE = $\alpha X + b$		CTE of refractory	
	a	b	Literature	Calculated
$ZrSiO_4$	−1.94	298	104	126
ZrO_2	−1.16	297	181	300
TiO_2	−0.69	300	229	240
Feldspar	−0.59	298	239	161

The CTE of the ground coat that was calculated from the linear regression equations (Table II) is the same as the directly measured one. That again, raises the question if there is dissolution of refractory in glass or if this dissolution does not affect the CTE.

We apply three different models to calculate the change in CTE with 10 vol% of refractory in enamel (Table III). The first model uses the Winkelmann and Schott (W&S) factors supplemented by Mayer *et al.*[5] The second model is based on Appen mol. factors.[4] Both calculations assume the total dissolution of refractory in glass. The third calculation is an average of the individual CTE proportionally to the volume content of the phases.

Table III shows that all three calculations give results similar to the experimental ones. That means CTE data are not reliable evidence for or against the dissolution of refractory in glass.

Feldspar decreases the CTE of enamel to a lesser extent than other refractories. A possible explanation is the presence of alkali metals in feldspar, which increases glass CTE far more than do other oxides.

Feldspar should be recommended when enamel is to be hardened without an increase in the compressive stress, as in case of spalling problems.

Table III. Changes in CTE with 10 vol% of Refractory in Glass

	TiO_2	ZrO_2	$ZrSiO_4$	Feldspar	SiO_2
CTE	240	300	126	161	15
Density (g/cm^3)	4.26	5.89	4.56	2.6	2.6
W&S	14	−6	−10	−1	−19
Appen	−17	0	−7	−7	−19
Average CTE	−3	0	−10	−12	−29
Experiment	−4	−5	−11	−5	

Conclusions

- Refractory additions, while they make enamel harder, do not change T_g. The traditional explanation why the harder enamel is more resistant to hairlines is not applicable to heterogeneous enamels.
- The thermal expansion data have been successfully used for years to predict frit performance. There is a good correlation between frit CTE and resistance to crazing, hairline, and spalling of enamel. As for multiphase enamel, the correlation is more complicated, and other factors should be taken into account.
- The CTE of individual phases of enamel alone are not a reliable factor to predict enamel thermal expansion and the stress in enamel. Glass interaction with refractory, bubble structure, and other factors substantially influence the enamel CTE.
- Stress-related enamel behavior correlates with various enamel properties. All these properties are related to enamel composition. In some cases, enamel behavior could be better explained by other than CTE properties. In these cases, correlation of this behavior with CTE just reflects common relation to the enamel composition.
- There is no polymorphic inversion in mill-added quartz during the firing cycle.
- Of the studied additions, feldspar is the least and zircon is the most efficient in decreasing enamel thermal expansion.
- Thermal expansion of enamel linearly correlates with the amount of refractory added.

References

1. F. Birch and P. LeCompte, "Temperature–Pressure Plane for Albine Composition," *Am. J. Sci.*, **258**, 209–17 (1960).
2. S. Pollington et al., "Determination of Veneering Ceramic for Fluocanasite Core Ceramic," presented at LADR/AADR/CADR 82 General Session, March 10–12, 2004.
4. W. D. Kingery, H. K. Bowen, and D. R. Uhlmann, *Introduction to Ceramics;* p. 604. Wiley, New York, 1976.
5. V. V. Vargin, *Technology of Enamels;* pp. 41–45. Maclaren and Sons, London, U.K., 1967.
6. B. J. Sweo and D. J. Snow, "Factors which Influence Hairlining of Enamel," *Proc. PEI,* **32**, 39–47 (1970).
7. A. I. Andrews, *Porcelain Enamels,* 2nd Ed.; pp. 58–78. Garrard Press, Champaign, Ill., 1961.

Changing the Rules— The Survival of American Manufacturing Depends on Everyone

Cullen L. Hackler
Porcelain Enamel Institute, Inc.

Introduction

Most American businesses have felt a significant impact from offshore manufacturers and imported products. In fact, 2003 established an all-time record trade deficit of $1.5 trillion. I believe we will have to "change the rules" to begin to reverse this negative trend. This will not be an easy task, and it is going to require the efforts of everyone.

Although I am not an economist and my educational background is ceramic engineering, I have more than 31 years of business experience, 26 of those with manufacturing companies. My comments today are an outgrowth of my role with the Porcelain Enamel Institute (PEI), where I have the opportunity to regularly discuss business issues with a variety of large and small manufacturers. PEI was founded in 1930 and, for almost 75 years, it has served its membership plus the industry as a whole. In our industry, some have already felt the crippling effects of offshore competition, some are competing head-to-head as we speak, some have domestic plus foreign operations (so they continue to struggle to find the proper balance), and some are feeling "safe" today but keep a wary eye on the future.

The manufacturing/economy issue is complex; we have heard it in the media, in industry conferences, and even from the U.S. Department of Commerce. I am convinced that every one of us will have a role to play as we work to reverse many of the negatives U.S. manufacturers face. Having said it is complex, I will now try to organize and simplify it a bit.

Overview

Here is an overview of what I want to discuss with you today. If I think about how things might look in a few years, I realize that a lot of things are at stake—wealth on this planet is not uniformly distributed, and, as Americans, we have more than anyone else. Other countries and economies are seeking to emulate our success; they are striving to approach or even exceed our standard of living. Let us face the reality: we, as Americans, are not universally loved around the world. Many of the inhabitants of this planet resent our status and affluence to the point, in some areas, of hatred. To quote someone far wiser than me, "… nothing strengthens the resolve of the weak as their hatred of the strong." Other countries are not going to worry about our problems, they are too busy and occupied working on their own issues in ways that significantly impact us.

These changes, if left to run their apparent current course will potentially culminate in huge alterations of our American way of life. I will review a number of the important issues that my member companies, as well as many other industries and manufacturers, face and that will lead into a look into some of the potential changes we might make. I will conclude with an emphasis on how we will all have to contribute to making real, lasting changes.

What is at Stake?

Over the past 15–20 years, we have seen ever-increasing trade deficits (2003 was a record) and a steady flow of American jobs to foreign manufacturers. Current statistics report a loss of more than 2.7 million manufacturing jobs, and economists estimate that almost a half a trillion dollars per year of American wealth flow offshore! In my view, an unfortunate situation has occurred; that is, the great strength and resiliency of the U.S. economy, plus some governmental financial policies, have tended to mask the negative impact of these lost jobs and continuing flow of money to other countries and economies.

How much longer can our economy absorb these "body punches" before we all begin to see a general loss in our own quality of life and standard of living? Most foreign manufacturers and countries do not have America's best interest at heart; they are working to improve their own way of life, not necessarily at the expense of ours, but it seems to be heading in that direction. It is the American dream, our way of life, that is at risk here at the dawn of the 21st century.

Jobs and Our Economy

Our economy is linked to manufacturing, and the health of our manufacturers depends on consumers with money to spend. I do not believe that the flight of jobs offshore is due to problems with the American worker or our manufacturing plants; if there were issues, then, for example, why have so many foreign car manufacturers built plants in the United States (Tennessee, Kentucky, Ohio, Alabama, South Carolina, etc.)? The problem is in the total system—it is not isolated with the manufacturer.

Henry Ford was right when he said, "... I want to pay my workers enough to buy my cars ... ." His workers made products and earned wages, and then his workers bought products with their wages. Thereby we have an economy that works—I know it is a simplistic view, but it all fits together. Our economy is, in reality, quite complex, and we will need to have all of the many pieces and players working in concert to turn things around. And, yes, the world economy is even more complex, and, yes, we are a big part of that economy—but we will have to start fixing the things we control. It is imperative that we get to work now on our own economy and issues. Manufacturing and manufacturing workers earning money are key pieces in the puzzle but not the only pieces.

Issues

Here are a number of issues that manufacturers face; this list follows from discussions with my member companies at our 2003 annual membership meeting—following the list is a discussion of each, in some detail.

- Low-cost producer game—do not get trapped;
- Price conscious consumers—it is almost all we hear in ads;
- Fragmented retail markets—we are beyond segmentation to fragmentation;
- Forgotten small businesses—large companies must remember their neighbors;
- Playing field is unfair—tariffs, laws, and infrastructures vary from country to country; and
- Management and labor must cooperate—we need to recognize the big picture.

Low-Cost Producer Game

As I mentioned before, the problem does not rest only with the manufacturer. I sincerely believe that the entire system, from raw materials in the ground to consumer products on the retail shelf, contributes to our current situation. Here is a scenario that I have, in general, seen over the past 15 years, and it is but one example of why we find ourselves having to deal with these problems today.

Rather than spend R&D dollars to ensure that their business adds value to the consumer/end product being produced, many manufacturing businesses have become too focused on minimizing the cost they add to a product. They then are forced into the "low-cost producer game" just to survive, but many evolve away from manufacturing into functions they think they can control, such as product design, assembly, and marketing—and they then continue to outsource more and more manufactured components. Unfortunately, they begin to add far less value than the actual manufacturer. They now face a further obsolescence risk, because the downstream distribution segment of the channel can completely skip them and deal directly with the actual manufacturer; realizing the risk, they continue to try to force costs down with pressure on their suppliers (who can fall into the same trap) and production units (increase productivity, decrease labor content, etc.) plus the elimination of all development, R&D, or innovation.

We now have totally changed the business. What was once a company with laboratories developing products and processes to solve consumer needs, now builds plants to produce those products and develops a distribution system to get the goods to the end users; this has become an altogether different animal. Our business has changed. Suppliers are expected to do the R&D or innovation in materials and manufacturing processing to meet design specifications. Actual production is outsourced, often by an RFQ via the Internet, and awarded to the lowest bidder. The distribution channel is controlled by the retailer, who asks for financial help to underwrite advertising. We (manufacturers) are just one piece in the overall chain, and we continue to add less and less value.

Do not misunderstand me, manufacturers have to improve productivity and efficiency and to eliminate waste. Being cost competitive is good whether your competition is across the street or halfway around the world. My point is that the entire answer does not rest solely on the shoulders of the manufacturer.

Price Conscious Consumers

Almost every TV ad we see from major retailers pounds us with "the guaranteed lowest price." Quality may or may not be mentioned. In fact, many brands are built to condition consumers to shop on price. Yet, I wonder, did the consumer demand cheaper prices from the retailer or have retailers chosen to compete with one another on price alone? I am convinced it is the latter, and I wonder if the consumer really is deriving any long-term benefit.

In order to offer that lowest price, Wal-Mart (and I am not picking on anyone here, just one example among many) has evolved into importing the vast majority of the goods it sells. They used to promote "made in America"; now they import most of what they sell, and we, as consumers, continue to buy. I think we consumers need to be reminded of the fact that we are all a part of the economy that supports our American manufacturers. Back to Henry Ford, we need to be sure that we buy the products our neighbors are manufacturing to ensure that they will earn wages and be able to buy what we provide. We may have to pay a little more, but at least it will be an investment in America, not in a foreign country.

Retail Markets are Fragmented

I have mentioned this before. As markets become not segmented and further fragmented, retailers are able to exert a lot of control over prices. In many industries, manufacturers rarely control the retail price of their products. Who is going to "change these rules"?

- A creative manufacturer (who uses technology to make a "better mouse trap");
- A distributor who is tired of the squeeze ("changes" the distribution channel);
- A retailer looking for a point of differentiation (i.e., "made in America"); or
- All of us as consumers (we only buy goods made by our neighbors).

None of these will be easy, and any and all of them will be risky, but something or someone is going to have to get us off our current slippery slope and change the direction.

Remember American Small Business

Building on the issue of buying from those who will buy from us, large U.S. manufacturers need to source products and services from the many small, local manufacturers ready to supply them. We all need to remember how the pieces fit together in our economy and take a long-term thinking approach rather than just making "cost-driven" outsourcing decisions.

Many large manufacturers have used technology to try to remain competitive; this is a very positive sign for American business. Many of these positive results and success stories represent large companies as excellent role models; they must become good big brothers, and they must share those technological advances with many small businesses that do not necessarily have the resources to have made such advances.

Not to be a broken record, but we all need to help those who can help us, really help us—this will require significant changes in how we do business and necessitate longer-term business thinking, financial objectives, and shareholder expectations.

The Playing Field is not Level

Government also can play a significant role in helping turn things around. Most of our manufacturers are happy to compete with anyone in the world as long as there is a level playing field (i.e., all are playing by the same rules). However, the impact of tariffs, dumping, and environmental compliance costs need to be equal for all, or we will continue to ask American manufacturers to fight with "one hand tied behind their backs!" Government must be sure to fulfill its responsibility to all Americans. The U.S. government, i.e., the legislative and executive branches, must exercise their authority to support our systems and maintain our own economic development, standard of living, and way of life.

Business–Worker Cooperation

Management and labor must cooperate to support our competitiveness. Both must work together to use technology and increase productivity, all in an overall effort to add more value to the products we manufacture. No one ever saved themselves to prosperity!

This cooperation will ensure job security for workers and competitive products for businesses; American labor cannot demand too much, because foreign labor is cheaper and the labor content of many manufactured products is significant. The use of new technology to improve productivity has to be a priority for business and workers alike, because it helps the enterprise be more competitive. This is just what our overall economy needs: businesses selling products and paying wages, workers earning money and buying other products.

Potential

We have huge potential for all aspects of our economy to participate. Management and workers must cooperate to ensure future employment and competitive products. Large and small businesses must work together. Government officials must stand firm to lobbying efforts and peer pressure to protect our jobs and keep them on American soil. Retailers must focus on giving consumers value/quality not just lowest price, and they must partner with American manufacturers to give the consumer "made in America" choices. And, finally, American consumers must be willing to pay more to purchase products made at home—keep your neighbors working so they can buy your product/service.

We Are in This Together

Everyone has a part to play, because we are linked in the total economy: business needs to recognize their link to the overall U.S. economy and make outsourcing decisions in that context; government must help support keeping jobs here; universities, technology consultants, and financial institutions must help lead and drive innovation; and consumers need to (1) understand the global picture, (2) recognize the consequences of foreign imports, and (3) develop a willingness to pay more to keep Americans working. Currently, this complex agenda lacks a leader; we critically need one to get all the pieces focused on the objective, but, until then, we will all have to find small, individual ways to work together and play our parts.

Conclusion

The early years of the 21st century likely will be a defining time in the history of the United States—we face big challenges, and, yet, we bring a legacy of entrepreneurship, innovation, and hard work, which, I believe, will see us through these difficult times. Although it is true that the genie is out of the bottle, and I agree that it will certainly be hard to put it back in, there is no doubt in my mind that we have the tools for success. However, I sincerely believe we can no longer afford to wait; our success depends on getting started today!

Notes

Notes

Notes

Notes

Printed and bound by CPI Group (UK) Ltd, Croydon, CR0 4YY

16/04/2025

14658456-0001